The five methods

Douglas Scarrett

Formerly Principal Lecturer in Valuations
Leicester Polytechnic

E & FN SPON
An Imprint of Routledge
London and New York

First published 1991 by E & FN Spon, an imprint of Chapman & Hall
Reprinted 1995, 1996

Reprinted 1998 by E & FN Spon, an imprint of Routledge
11 New Fetter Lane, London EC4P 4EE
29 West 35th Street, New York, NY 10001

© 1991 Douglas Scarrett

Typeset in 10/12pt Times by Best-set Typesetter Ltd, Hong Kong
Printed and bound in Great Britain by Page Brothers, Norwich

British Library Cataloguing in Publication Data
A catalogue record for this book is available from the British Library

Library of Congress Cataloguing in Publication Data
A catalogue record for this book is available from the Library of Congress

ISBN 0–419–13780–7

∞ Printed on acid-free text paper, manufactured in accordance with ANSI/NISO
Z39.48-1992 and ANSI/NISO Z39.48-1984 (Permanence of Paper)

Contents

Preface

The five methods of valuation are well established and between them provide the basis for valuations for a wide range of purposes.

The measurement of value should be based on a sound knowledge of the method or methods used. The book has therefore taken each method in turn and explored its use in practice. In this context, the judges and members of the Lands Tribunal have played an important part in developing an understanding of the thought processes involved, to which references have been made.

Throughout, the emphasis is on understanding the approach, more reliance being placed on explanation than on numerical examples.

Some attempt has been made to describe recent developments in the approach to valuation which it is hoped will be useful to practitioners as well as students.

It is my pleasant task to thank all those who have contributed to the contents of this book, wittingly or otherwise; my especial gratitude is reserved for Beryl and Emma.

The current domestic rating system

The community charge was replaced by a new system as from 1 April 1993 and comprises a combined property and personal tax which represents the local contribution to funds for services provided by local authorities and the police authority.

Each separate housing unit is placed in one of eight bands according to its estimated value at 1 April 1991. The bands are:

A up to and including	£40000
B	£40001 to £52000
C	£52001 to £68000
D	£68001 to £88000
E	£88001 to £120000
F	£120001 to £160000
G	£160001 to £320000
H	more than £320000

Although the rate payable is determined primarily by the band into which the dwelling falls, a personal element is introduced in the form of discounts for, among others:-

full time students including student nurses, apprentices and Youth Training trainees
certain other young persons
patients in hospital
people being looked after in care houses, those who are severely mentally impaired or those staying in certain hostels and night shelters
members of religious communities
people in prison (except those held for non payment of council tax or a fine)

The full tax is payable on the assumption that two adults reside in the dwelling. Where only one adult occupies the unit, the council tax is reduced by a quarter, where the dwelling is no-one's main home, council tax is payable at the rate of 50% (this applies largely to second homes and to empty dwellings, the latter being exempt from charge for up to six months provided they are unfurnished).

There are other exemptions associated with the state of repair or the status of the occupier (e.g. trustee in bankruptcy, dwelling occupied only by persons under 18 years of age or who are severely mentally impaired, repossessed dwellings or dwellings waiting to be occupied by a minister of religion).

Appeals against a particular banding are allowed only in limited cases

— where there has been a material increase or reduction in value from works carried out to the dwelling (although any upward revaluation would not take effect until after a sale)

— a new owner may appeal within six months of acquisition (unless the same appeal has already been made and determined).

Any appeal will be considered by the local Valuation Tribunal.

LEASE TERMS – RENT REVIEW

There is a reference on page 62 to relitigation of an important point related to rent reviews. As a result of a later hearing, the tenants have been denied the opportunity to test the matter again. The court held that section 1(7)(b) of the Arbitration Act 1979 allowed only one application and that the application had to be made within the time limit imposed. (*National Westminster Bank plc v. Arthur Young McClelland Moores & Co (1990) 50 EG 45*).

D.S.
September 1996

The background to valuation

PART ONE

Setting the scene

Ever since man began to foresake his nomadic existence in favour of permanent settlements, possession and ownership have been the concern of society. Land use patterns and occupation rights demand a set of rules to avoid the necessity to defend title by remaining physically in possession at all times. As society matured, a legal system evolved providing the framework with which such matters were settled thereby avoiding the alternative of trial by strength.

The evolution of the regulatory framework was probably preceded by economic activity on a barter basis. But even the most rudimentary exchange system requires measurement and assessment, so skills developed in relation to land at an early stage in history. The English legal system has developed from the feudal order imposed following the Norman conquest. Thus the Domesday book completed in 1086 recorded the property values, rights and obligations of the King, his tenants in chief, their subtenants and the peasants, whilst a court roll contained similar information for each manorial estate.

An Act of Parliament as early as 1695 laid down the rental value multiplier to be used as the basis for compensation when weirs were pulled down to carry out improvements to natural rivers to make them navigable for inland transportation — the forerunners of specially constructed canals.

Until some time after the end of the Second World War, the property market was to a large extent localized, low key and fairly predictable, with much investment property occupied for the operational use of companies controlled by the investing owner or in which he had a financial interest. The property investment scene has changed out of all recognition since those times.

The immediate problem in 1945 was the reinstatement of cities devastated by war and substantial building activity in places like, for example, Bristol and Coventry, resulted in the establishment of completely new and relocated shopping centres. At the same time, the provision of housing was regarded as a public sector priority with the local authorities being encouraged to take a role as major providers. To this end the phrase 'working classes' used in the Housing Act 1936 was dropped from the 1949 Act of the same name. Even including the contribution by the private sector, central government found difficulty in approaching its target of 500 000 units a year.

Meanwhile, the Town and Country Planning Act of 1947 made a bold but ultimately unsuccessful attempt to deal with improvement by providing for the expropriation of private development rights in return for a claim on a 'once and for all' notional global sum of compensation of £300 000 000.

A rigid code of land use was introduced in 1948, giving planning a largely restrictive role in the development process and exerting a major influence on land values, ensuring widespread use of the planning appeal procedure.

The private sector recognized the potential for multiplying the value of suitable land by developing it with low cost, non-equity capital readily supplied by the banks and others, often to the full extent of the development costs, confident of their security from the anticipated value of the completed scheme.

The local authority structure was unsuited, and its financial ability unequal, to the task of commercial development, and the pressing need to foster development to create job opportunities was not then an issue. Many authorities had no wish to become involved and often disposed of land carefully pieced together, through the use of compulsory powers, to commercial groups anxious to exploit the development potential.

It was not until the mid 1960s that insurance companies began to look for a share in the equity of schemes for which they had provided much of the venture capital to the increasingly large national property investment companies. Various forms of sharing were devised, a popular one being the provision of funds at rates below market levels in exchange for a share in the equity or even for the purchase of the whole scheme based on an agreed formula related to the final rent roll, once it had been completed and let successfully. In such cases the developer acted as a catalyst, using his expertise in the development process in return for a capital profit — a form of entrepreneurial project manager.

It is difficult to understand now how much the activities of a few developers, for example, Levy, Hyams, Clore and Cotton dominated the news pages of even the popular press. They were variously acclaimed as financial wizards or denigrated as parasites depending on the observer's point of view. Undoubtedly they had flair but they were gradually overtaken as schemes became larger and property companies adopted more formalized structures and became more accountable to their shareholders.

Market activity continued and intensified to a point where — inevitably it would seem with hindsight — there was a spectacular market crash in the early 1970s. Some depositors anticipated the seriousness of the situation and hastened the crisis by transferring cash to the larger banks in 'a fit of collective prudence'. The crash affected not only the property companies but extended to the whole of the secondary banking system to such an extent that the Bank of England intervened with a financial lifeboat to mitigate the worst effects of the substantial losses suffered. The liability was shared with the major clearing banks but the details were, and are, shrouded in secrecy. It was some time before confidence returned to the property market.

Eventually it did so, with the players becoming increasingly ambitious and pursuing imaginative town centre schemes, office developments on a vast scale and out-of-town business and shopping parks. On any dispassionate view, the over provision of retail and office space in particular is likely to lead to some bitter marketing and rental competition, resulting in greatly accelerated depreciation of the schemes losing out.

The day of reckoning may be nearer than many would wish to acknowledge, hastened by a base rate of 15% with an official rate of inflation of over 10%. There has also been a redistribution of the rate burden, brought about by the introduction of the National Business Rate and bearing heavily on certain types of user, in particular retailers and office occupiers.

Should the result be a major crisis in property development in the early 1990s, it will be the second painful lesson within 20 years. But the market is now more ordered and better structured than in the 1970s and the institutions are much better prepared to weather the storm and absorb the losses than were the earlier investment companies.

It is against this uncertain background that the principles of the various methods of valuation have been described. The work of various commentators has been noted but the underlying principle has been first to place property investment in context in the overall investment market and then to give a detailed discussion of the different approaches, leaving the valuer to decide the extent to which he wishes to adapt his practice to produce the required valuation. Such decisions will be taken against a background of great structural change in society and advances made in the research and analysis of non-property investments. Whilst the view that techniques that have stood the test of time should not be abandoned lightly is well understood, there is a need to respond to the needs of clients for increasingly detailed information and analysis in support of the resulting valuation.

The increasing sophistication and internationalization of the overall investment market suggests that unless the valuation profession moves to improve the quality of its general advice and market analysis the work will gradually pass to others; indeed it has already done so to some extent.

2 | The general investment market

Funds for investment by companies and for the massive expenditure incurred by the state are financed in part by the public, sometimes directly but more often indirectly. Very few private investors have access to sufficient funds to enable them to acquire substantial amounts of stocks and shares, but indirectly their individually modest contributions to pension funds and insurance policies provide vast sums. The financial institutions receiving these payments wield immense power in the investment field.

As personal wealth and influence have been replaced by institutional funds it has brought an increasingly sophisticated and knowledgeable approach to the art of investment. Institutional investors tend to manage their funds in an active way, being prepared to switch from one holding to another when demanded by market indications. The fund manager bases his decisions on market research into performance and forecasts prepared by analysts. Institutions act conservatively with regard to risk taking. They avoid the more volatile stock and attempt to build and maintain a balanced portfolio — a range of investments forming a complementary whole, spreading their exposure to risk. More recently, international investment has taken on a greater significance.

Investments may range from government stock to shares in companies, from debentures and preference shares to tenanted buildings. The individual may invest or save in a modest way by depositing amounts in a building society or bank, by contributing to a pension fund or by taking out an endowment policy to mature at some future date. Many investors of modest means have been encouraged to buy shares in newly privatized companies though some have been tempted to sell and take an early profit.

The various ways in which funds contributed by an individual are channelled are shown in Figure 2.1. The individual not only finds it beyond his knowledge and financial ability to invest significantly in specific com-

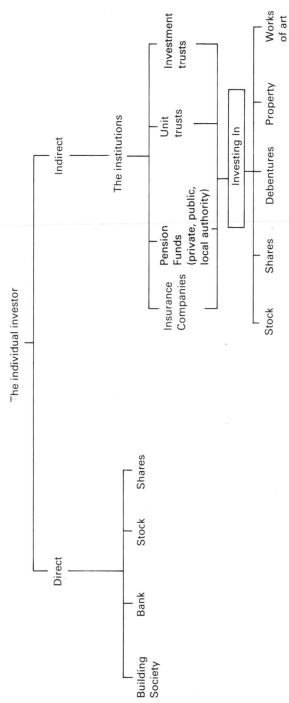

Figure 2.1 Direct and indirect investment by the individual.

panies but it may be more tax efficient to invest indirectly. He also avoids the exposure to risk that investment in a very few companies would involve and he receives the benefit, indirectly, of professional expertise and management.

2.1 TYPES OF INVESTMENT

The investor has a range of opportunities, each with its own characteristics. The main difference is between fixed and variable income producing investments.

2.1.1 Fixed interest investments

The first category includes government stock, company loan stocks and preference shares.

Government stock
Government stock is issued by the government to finance its massive expenditure. It is referred to as 'gilt edged', reflecting the underlying security afforded by the backing of the state. It is inconceivable that the government would default in interest payments or on redemption, giving the stock a security status against which all other forms of investment tend to be measured.

   Government stock is usually issued at par of £100 nominal value although the initial price may be above or below the par value which will affect the yield offered. The yield will be paid on the par value so the effective yield must be calculated where the initial or subsequent selling price is other than par. Most issues have a date which refers to the year when the stock will be redeemed at its par value, so even a stock with a low interest yield will rise in value as it approaches its redemption date, as illustrated below.

Example 2.1

A Treasury stock with a redemption date of 2003 and an interest rate of 8.25% is issued at £96 for each £100 of stock. The running yield taking into account the payment of the 'coupon' rate of interest receivable on the issue price of the stock must take account of the lower purchase price.
   Thus, the running yield is

$$\frac{\text{Par value}}{\text{Price}} \times \text{coupon}$$

$$= \frac{100}{84} \times 8.25\%$$

$$= 9.82\%$$

The investor can now compare this return with other interest rates available in the market.

Where the stock is issued at a price in excess of its par value, the effective interest rate can be calculated in a similar way, given the issue price and rate of interest.

Example 2.2

Calculate the yield on Exchequer 1998 9.5% issued at £105.
   The running yield is

$$\frac{\text{Par value}}{\text{Price}} \times \text{coupon}$$

$$= \frac{100}{105} \times 9.5\%$$

$$= 9.05\%$$

An investor wishing to buy or sell stock during its life will not deal with the government but will sell through the stock exchange. The market will fix the price according to the yield attached to the stock and its perception of current yields and levels.

Example 2.3

Calculate the running yield on undated 3.5% War Loan at a price of £34.5.
   The running yield is

$$\frac{100}{34.5} \times 3.5 = 10.14\%$$

In other words, the investor can buy £100 stock in the market for £34.50, the adjustment of the purchase price assuring a more acceptable return than that fixed when the stock was issued many years ago, since the coupon rate of 3.5% is paid on each unit of stock, even though purchased at a price well below par.

In the case of dated stock, redemption will take place in the year indicated when the par value will be received.

Example 2.4

Treasury 8.5% 1994 is quoted at a price of £89 and shows an interest only yield of 9.5%. The gross redemption yield is approximately 12.05% reflecting the fact that the stock will be redeemed in 1994 at its par value (both figures are available from the financial pages of *The Times* and other daily newspapers).

Stocks are categorized as follows:

Short-dated (shorts)	redeemable within 5 years
Medium-dated (medium)	redeemable 5–15 years
Long-dated (longs)	redeemable after 15 years
One way option stocks	(undated or irredeemable)

The option category comprises stock redeemable only at the option of the government. Some have no date for redemption whilst others have a date such as '19.. or after' Where such stocks are coupled with a low rate of interest, it is unlikely that the government will exercise its option to redeem and trading in the market will ignore the possibility of redemption. More recently the government has introduced index-linked stock which is described here because it is another form of government stock although not carrying a fixed interest rate. A sum is invested and receives an interest payment varying in line with the Retail Price Index (RPI). It has a fixed life at the end of which the nominal value is repaid in full plus the RPI increase. Both interest and capital are index linked.

Other fixed interest securities
Other issues include dated and undated stock with interest at various rates payable quarterly and half-yearly. The issues tend to be small and specific and consequently trading is less active. Such stocks are at a disadvantage since capital gains tax is payable on them but not on British Government Stock.

Local authorities issued negotiable short-term bonds at one time, often for as little as a year (yearling bonds).

Some public boards raise money on the loan market although with privatization of water, gas and other services, the market is smaller than before.

Loan stock is issued by companies as one form of share capital, secured on particular assets of the company and rewarded with a fixed rate of interest. Such investors have no other rights in the company.

A debenture stock is usually secured on specific assets of the company with a further floating charge over the remaining assets of the company. The advantage of debenture stock is its prior ranking for payment if the company is wound up.

2.1.2 Variable interest investments

The second category includes company shares (equities), unit and investment trusts. The return is not fixed or guaranteed.

Equities
The holders of ordinary shares in companies are the risk takers. When a company is prosperous, the share dividend is likely to increase; should

the company run into difficulties the shareholders may receive a much smaller dividend or no dividend at all; should the company collapse the shareholders may find that their investment is worthless.

The gross dividend yield is calculated after the net of tax dividend is grossed up to make the return comparable with other returns. The formula is

$$GDY = \frac{D}{P} \times \frac{100}{(100-T)}\%$$

where   GDY = gross dividend yield
$\quad\quad$ D = dividend
$\quad\quad$ P = price
$\quad\quad$ T = tax rate

Example 2.5

Where a share whose price is £6.52 attracts a net dividend of £0.45 and the tax rate is 40% the gross return or yield is

$$\frac{0.45}{6.52} \times \frac{100}{60} = 11.5\%$$

The earnings yield is sometimes quoted and illustrates the ability of the company to pay since it is based on the total surplus as opposed to the amount allocated for dividends which may be reduced by a substantial transfer of earnings to reserves. It is calculated from

$$Earnings\ yield = \frac{earnings\ per\ share}{market\ price\ of\ share} \times 100$$

or may be calculated by multiplying the dividend yield by the dividend cover. Another way of expressing the yield is to calculate the price/ earnings (P/E) ratio which is the reciprocal of the earnings yield. Preference shareholders are repaid after creditors but before ordinary shareholders on the company being wound up. Preference shares receive more favourable tax treatment than loan stock. Different issues of preference shares may be ranked or treated on an equal basis. Most preference shares are cumulative which means that a dividend passed in one year is carried forward until it is paid. A non-cumulative preference share loses the payment for any year in which the dividend is passed.

Convertible debentures or loan stock attract fixed interest payments but carry rights to convert to ordinary shares at the holder's option subject to the terms of issue.

Unit trusts
For the investor without the financial resources to buy and hold a range of shares over a wide spectrum and or without the time and expertise to

devote to their management, unit trusts offer an alternative. Each trust consists of a form of managed fund safeguarded by a trustee, invariably a bank or insurance company, which holds the assets, controls the issue of units, maintains a register of holders and oversees the general management. The investor provides funds which are applied to the purchase of stocks and shares. He can require his share to be repaid in cash at any time although the amount is not guaranteed but is based on the value of the fund at the time of sale. The trust will have declared aims and may specialize in certain sectors of the market or certain features, for example, it may be an income fund, a growth fund, a balanced fund or a specialized fund; the terms emphasizing the thrust of the investment policy pursued. The trust has no separate capital structure and ultimately all the funds including income belong to the unit holders.

Investment trusts
Despite their title, investment trusts are limited liability companies having share capital and the right to issue prior charge capital and to raise new money by rights issues.

Inland Revenue approval to their operation is needed if they are to enjoy the tax advantages usually associated with such trusts. Most of the company's income must come from securities of which only a limited amount may be retained each year. There is also a limitation on how large a proportion of the funds may be invested in a particular holding. Capital profits must not be distributed. The investment trust should avoid excessive dealing since this may endanger its status.

Business expansion schemes (BES)
The original aim of the government in promoting the business expansion scheme was to facilitate equity investment by individuals in unquoted companies engaged in manufacturing and general commercial activity. In return, there are valuable tax incentives at the investor's marginal rate, combined with exemptions from capital gains. Both benefits are subject to minimum holding periods.

The scheme attracted £105 000 000 in 1983–1984, the first year of operation, of which about £20 000 000 was devoted to the purchase of agricultural land, an outlet later withdrawn. In the following year, property development attracted £46 000 000 with further investment in other property activities. Property development was then excluded from the scheme, only to be reinstated, with farming, as qualifying trades subject to interests in land and buildings accounting for no more than 50% of a company's net assets subject to a funding ceiling of £500 000 in any 12 month period. The 50% limit was overcome in many cases by matching property assets with loan charges, and this has been accepted by the Inland Revenue.

Following the 1988 Budget £370 000 000 (well over 90% of total BES funds for the year) was invested in companies providing residential accommodation under the less restrictive provisions of the Housing Act 1988, which removed rent ceilings. In this case the rule limiting the interest in land and buildings to 50% does not apply and the funding limit per company is fixed at the higher figure of £5 000 000. The cost of individual properties is restricted to £85 000 (including furnishings where applicable) and £125 000 in Greater London.

In general the scheme offers additional opportunities to investors but subject to some element of risk. For the government there is the trade off of signalling its willingness to recognize particular investment objectives and offering tax benefits to those investors prepared to support the identified areas.

2.1.3 Other investment media

The other main avenues of investment are property and items expected to have potential for capital appreciation.

Property
The investor may purchase freehold or leasehold interests in properties which are let for business, commercial, industrial, residential, leisure or agricultural use. The investor may be entitled to the rents issuing out of occupation of the land or buildings or simply to the ground rents where another investor erects and lets the building. Ground rents are particularly well secured and offer attractive investment opportunities provided the owner shares in the prosperity of the built scheme through regular rent reviews.

Leasehold interests are wasting assets, resulting in the cessation of income and a nil capital value at the end of the term. Terminal dilapidation claims may be onerous and the ground lease should be studied carefully before a commitment to purchase is made, particularly where the lease has a short unexpired term.

The large sums involved in the acquisition of property investment limit the ability of the single investor to have a sufficient holding to achieve diversity of risk, ensuring that non-operational property is held for the most part only by large-scale investors.

Becaue most property is let on upward-only rent reviews — where the rent on review cannot drop below the present level even where there has been no increase or where there has been a decrease in rental value — the investor is assured of a basic return during the term of the lease. However, where the property has reached the end of its useful life, the investor may experience difficulty in re-letting at the end of a lease term. Most institutions include a property element in their investment portfolio without allowing it to dominate. The fund managers are conscious of the drawbacks of property investment including illiquidity, large size and indivisibility of holding together with significant management costs. The smaller investor for whom direct investment in property is impracticable may invest indirectly through the purchase of property bonds or shares in property companies.

Capital appreciation

Investors without need of income are able to consider the purchase of items the capital value of which is expected to increase substantially. Works of art are sometimes purchased on this basis; the best-known example was that of the British Rail Pension Fund which purchased valuable paintings and other items which were sold recently at very satisfactory prices, amply compensating for the absence of any income and the cost of storage and security over the years. The purchase of vintage cars is particularly attractive since no Capital Gains Tax is incurred on sale (unless the investor is engaged in dealing).

Indirect investment

Most indirect investment takes place through contracts of life assurance (entered into to finance home purchase and provide life cover linked with lump sums on maturity of the policy) and occupational pension fund contributions.

In recent years, following the abolition of exchange controls in 1979, the investment market has become much more international. There has also been active interest from European and Japanese investors for the limited number of high quality investment properties on the market at any one time. There is some support for the weight of money hypothesis which suggests that given the volume of funds seeking a return, there will be a shortage of better quality investment which may lead to a price increase to a level having little justification on the basis of the return available.

The alternative available to investors is to retain their assets in cash or near cash. In times of high interest rates it may be that the interest on money deposits is greater than the return from stocks or shares. If there is appreciable inflation at the same time, the purchasing power of the liquid assets will be eroded; the real return is the difference between the yield and the rate of inflation.

2.2 YIELD AND RISK

The Bank of England fixes a minimum lending rate. At one level the action has a political role, sending signals to the market about the government's economic policy. It is also influential in determining other interest rates.

Yields are also influenced by the perceived risk of the particular investment. In general the price of higher yields is greater exposure to risk.

Fraser has defined investment risk as the potential annual variability of the internal rate of return outcome from the investment's expected internal rate of return. The internal rate of return is susceptible to changes in both income flow and capital value.

Figure 2.2 suggests a typical yield structure for a range of investment opportunities and makes broad comparisons with investment in property. The total yield is made up of a risk-free return and a risk premium. The risk-free return is taken at a figure slightly below the yield on government stock given that even the safest investment is uncertain and involves an element of risk (though not on default in the case of government stock). The figure indicates the yield expected on property investments and compares it with the yield on other, non-property, investments. In the case of property, the initial yield may be appreciably lower than what would appear to be an acceptable yield because it takes into account the fact

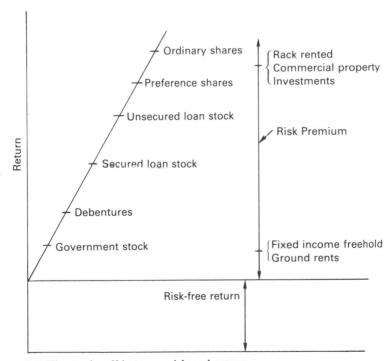

Figure 2.2 The trade off between risk and return.

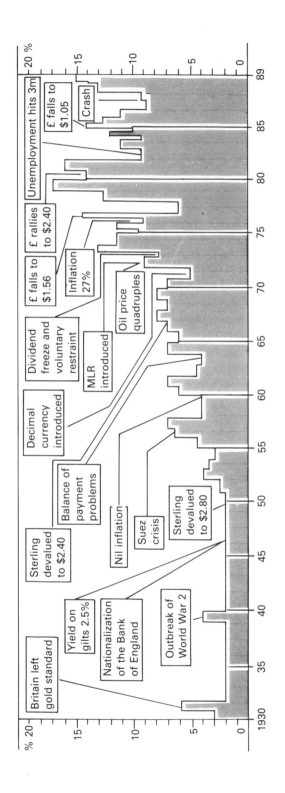

Figure 2.3 Clearing bank base rates showing trends and events over the last 60 years.

Source: *Investors Chronicle 10 November 1989*

that the property is let subject to frequent rent reviews and the expectation that the rent will increase at each review, due to rental growth.

The minimum lending rate has varied considerably especially since the end of the Second World War.

The relationship between United Kingdom clearing bank base rates and national and international events is shown in Figure 2.3.

3 The property market

Unlike stocks and shares, not all property is held for investment purposes. Much is held primarily for the operational purposes of a company and whilst the value may increase, the property will be important for the part it plays in enabling the company to be efficient and make profits. Occasions may arise where the value of the property to a third party will increase the value to such an extent that the directors may be persuaded to contemplate sale. Even then, the effect on the operation of the company will normally continue to be the primary consideration. Property held for the operational purposes of public or statutory bodies is even less likely to be offered for sale unless it is redundant or has been replaced. Nevertheless there are occasions on which a valuation may be required of such holdings or of development land, where planning permission creates value. One or more of the methods of valuation described, other than the investment method, may be appropriate in such a case.

3.1 PROPERTY AS AN INVESTMENT

The property investment market must be seen as part of the overall investment market whilst at the same time being a distinct segment having its unique attributes and peculiarities.

It is reasonable to suppose that any prospective investor will look at all the investment possibilities before deciding where to commit his funds. Most investors wish both to maximize annual returns and to protect the purchasing power of the capital invested. Consequently an investor may decide not to invest all his capital in any particular form of investment. For example, government stock and other fixed interest investments offer little or no capital growth. But the decision to move away from those areas of investment where a stock or share is acquired and a coupon or dividend received half-yearly to the ownership of a physical building is a major step requiring skills of financial and physical management and access to substantial funds.

In the past, part of the ethos of owning real estate was the prestige associated with this outward display of affluence. In ancient Greece, for example, property ownership was fundamental; it was the basis of a citizen's social status and political influence and the law went to great lengths to keep property within the family. Even in our own society, a woman's property became her husband's at marriage until the enactment of the Married Women's Property Acts of the late Victorian period. The right to vote was not finally separated from property ownership until the early part of this century.

An investor wishing to invest in equities may have his attention drawn to the shares issued by the major oil companies, the main chemical manufacturers or the leading retail chains. An investor in government stock has a wide choice of well documented issues and may choose between new issues or a purchase in the market, between short and long terms or even acquire undated stock. The same investor wishing to invest in property has a choice of developments used for industrial, commercial or retail purposes in addition to more recent activities such as leisure uses or may consider investing in agricultural land. His choice of a suitable investment within any of these sectors is far more subjective and less susceptible to advice derived directly from the market. He has to decide whether to restrict his search to the United Kingdom. If so, should he restrict his search to property investments in London, include the home counties or extend his enquiries to other parts of the country? He will need to judge the quality of the building and its life expectancy. If the building is already let, what is the standing of the tenant and is the lease well drafted? What are the opportunities for income growth and capital appreciation? All of these matters need answers which can only be provided by employing an adviser who is experienced and knowledgeable in the particular type of property under consideration. This advice comes at a price in addition to which there are substantial legal costs and an expensive stamp duty to pay on transfer. Once the property has been acquired there is the need to arrange for its management and the expense of maintaining it in sound structural and satisfactory decorative condition. The risk of fire and other disasters is normally laid off by taking out insurance cover, a further annual expense.

As noted elsewhere, investment in property has several unique features in addition to those shared with investments in general. As a result it could be argued that property investment demands additional skills and abilities to discern those desirable qualities and opportunities.

3.2 THE PERFECT INVESTMENT

The attributes of a perfect investment require that:

1. It is easily and speedily bought and sold, without restrictions on access and with low dealing and transfer costs;

2. It produces an income which compensates for the effect of inflation at least;
3. It is homogeneous and divisible;
4. It is well and fully documented with a large body of prospective buyers and sellers;
5. There is a demand for that investment;
6. It requires minimal management;
7. It is not likely to be the subject of government interference.

The absence of any of these qualities will not necessarily exclude the investment under consideration from selection. The investor may decide not to proceed with the purchase or to offer less to reflect a particular disability. He may even decide to look for other forms of investment which meet his requirements more fully.

At first sight, property investment does not fair well against the criteria listed, yet there is no shortage of willing investors. It is worth considering each of the requirements listed above and noting their effect in the case of property investment.

3.2.1 Ease of dealing

No one would claim that property can be bought and sold easily or cheaply. There may be considerable delay in completing the sale even where the investment is attractive to the market.

Full information has to be prepared and a marketing campaign undertaken and when an acceptable offer has been made there are legal processes to observe and possibly financial support to arrange. As a result, the transfer time of a property investment from one owner to another is likely to be measured in months in contrast to the fortnightly accounting period operating on the stock market. Such an investment is described as illiquid.

Not only are there lengthy periods of uncertainty before the transaction is finalized but the transfer costs involving professional fees and stamp duty are likely to be quite substantial. Costs of purchase and sale range between an estimated 2.75% and 4%; these figures include stamp duty of 1% on purchase price normally paid by the purchaser. Even abortive negotiations are likely to result in unrecoverable costs for both parties. It is generally accepted that property investment requires long-term ownership though this precept needs to be balanced with the needs of portfolio management.

3.2.2 Inflation

The effect of inflation is important. The ideal investment is one where the yield is guaranteed to at least equal the prevailing level of inflation. In a sense, any interest level below the level of inflation represents an unsatisfactory position, since the real interest rate is the difference between the actual return and the current level of inflation. Given recent high rates of inflation, the higher rates of interest on offer from government stock and deposits with building societies are put in perspective.

Investment in property has always been seen to offer good income growth; as a result, the initial yield is likely to be lower than that from fixed income investments to reflect the prospect of future increases. Any increase in annual returns tends to have a 'knock-on' effect in the form of increased capital values.

It should be noted that income in the form of rents is usually fixed for periods of 5 years, creating a lag in the process of maintaining an income to counter inflation. This is particularly acute where the owner retains the responsibility for repair and insurance where rising costs will erode the net income.

3.2.3 Homogeneity and divisibility

Each property is unique even if only because it stands on its own site. Even properties built to similar plans and with similar appearances may have significant differences. For example, two externally similar properties may be designed internally in quite different ways, and have planning permission for different uses.

An important area where property investment is at a disadvantage is in its indivisibility. If the investor wishes to raise money, he may have no alternative but to sell a property, even though this may produce a larger amount of capital than he needs which leads to the problem of reinvesting the balance at what may be an inconvenient time.

This is in contrast to the ownership of stocks and shares where a broker could be instructed to sell sufficient units to meet immediate requirements. It is true that the owner could investigate the possibility of arranging a short- or long-term loan against the security of the property but the cost may be uneconomic and the other terms unattractive.

Indivisibility is compounded by the relatively long and uncertain time scale affecting property transactions and the high costs of marketing, negotiation and transfer.

Some progress has been made in proposals to package individual properties for sale in small units, similar to shares. It is difficult to see how such shares could fail to trade at a discount to asset value. Also, such a holding would preclude an entrepreneurial management approach by individual shareholders.

3.2.4 Adequate information and a wide market

The property world is traditionally one of secrecy where the onlooker has only a limited opportunity to check claims or rumour relating to a particular transaction. Such secrecy makes complete analysis and market intelligence difficult to achieve. More information is published now than a few years ago but is necessarily selective; it is almost always a smoothed account of the market factor in general and may be misleading when attempts are made to relate it to the particular investment under consideration. Furthermore, the number of transactions is relatively small whilst the amounts involved are often very large. These features do not help interpretation of the market as the level of activity is often insufficient to suggest a pattern.

Plans have been announced recently for Land Registry information to be more readily and publicly available. It is not clear whether the information to be provided will include details of purchase price as is the case in Scotland; access to such detail would be a great step forward.

3.2.5 Demand

Economists point to the inelasticity of supply of property brought about by the extended time scale needed to respond to a perceived increase in demand. It may be possible for the owner to take advantage of an unfulfilled demand in the short term. In the longer term, the market is likely to respond but the provision is unorganized and fragmented and if that leads to an eventual over supply, the owner will find himself competing in a market where prices will tend to become depressed. The planning control mechanism is not primarily concerned with matching supply and demand.

3.2.6 Costs of management

In comparison with other investments, property involves a high cost of management. Once acquired, the property needs to be managed to ensure that rents are paid promptly and regularly, covenants are complied with and the property maintained in good condition. Management is usually entrusted to a firm of surveyors which will also advise on the level of insurance cover and deal with negotiations for increasing the rent on review and at the end of the lease. The overreaching aspects of management extend to responding to any occurrences likely to affect the client's property (e.g., a proposal to build an office block adjoining which may obscure valuable rights of light) and of pursuing any opportunities to enhance the value of the property (e.g., the possibility of a planning application to change the use to a more profitable one or of acquiring an interest in an adjoining property).

3.2.7 Government interference

The investor will be anxious that there is no likelihood of legislation or other government action to restrict the way in which the investment is managed or traded so as to affect the income return or capital value or to add to the costs of management. Property is owned within a statutory framework. The property statutes of the 1920s are important in regulating title and settling disputes. They are a great improvement on previous legislation, but are undeniably complicated and the resolution of any dispute through the courts is likely to be expensive and time consuming.

The majority of residential tenancies are subject to security of tenure for the tenant and to restriction on the level of rents charged. This position has been relaxed recently in relation to new tenancies but investors are likely to be wary of pure investment in rented housing given the history of substantial interference with contracts by ruling parties of all political persuasion dating back to 1915.

Agricultural rents can be revised at any time subject to the terms of the agreement between the parties or in the absence of an agreement at intervals of 3 years. Recent reports point to a low level of activity in the trading of tenanted farms, it being suggested that investors are discouraged by the security of tenure available not only to the present tenant for his lifetime but to members of his family. Further uncertainties are introduced by the commodity surpluses, the need to reduce the cost of maintaining the common agricultural policy, and the increasing politicization of farming issues.

It is not proposed to pursue the legislation relating to agricultural tenancies or the valuation of such investments which are outside the scope of this work.

Most commercial premises let on lease are subject to the provisions of the Landlord and Tenant Acts 1927 and 1954 and the Law of Property Act 1969. There is security of tenure except in certain specified cases and the tenant may claim compensation where the landlord establishes the right to possession. Compensation is also available to the tenant for improvement work carried out by him during the lease and the work may also result in a lower rent on gaining a new tenancy, subject to which a market rent is payable.

The parties are free to negotiate their own terms within the legal framework; any terms on renewal which the parties cannot agree may be referred to the courts by either party. Investors in general find comfort in a commitment by a tenant for a long term provided there is no artificial depression of rent or of the ability to review it at regular intervals. The acceptance of this legislation is witnessed by its longevity, almost unchanged, and the strength of and activity in the commercial property investment market. The majority of property investment is potentially within this legislation which is described in more detail later.

3.3 VALUATIONS FOR OTHER PURPOSES

In addition to the acquisition of property as an investment, it is often held for purposes where its market value is incidental and of secondary importance.

For example, the operational land of any enterprise, for example, British Rail, ICI and British Coal, enables the company to operate and generate income and hopefully profits. The total value of all the holdings for the purpose of that company is likely to be more than the sum of its parts. British Rail could not sell a section of its main London to Edinburgh line whatever the price offered. ICI would not sell one of its plants that performed one of the sequential operations necessary to refine oil for use in the plastics and petroleum industries unless it could maintain the continuity of operations.

The amount at which the assets are shown in the balance sheet is not reliable or conclusive as an indication of market value. It may even be misleading. Some companies write down all their properties to a nominal value whilst others inlcude them at cost but charge depreciation at the end of each financial year. Accounts including property assets valued by reference to the RICS Asset Valuation guidelines will not necessarily show the market values of the properties. Moreover, the directors are empowered to reduce the amount where they are not satisfied that the business would be profitable employing assets at the value given. Consequently, the valuer is required to assess values which are not necessarily market values and to value properties which rarely if ever appear on the market for disposal. Valuations may be required for compulsory acquisition, inheritance tax, transfer, exchange, division and other purposes.

Where there is a sizeable market of similar properties changing hands on a regular basis it may be possible to make a direct comparison of capital values, as in the case of dwellings for owner occupation or of hotels. Where the property is let and held for investment purposes, the valuer will compare the rents and other terms and also the yields acceptable to purchasers for different types of investment. These two processes describe the comparative and investment methods of valuation although it can be seen that the latter includes elements of comparison also. These are the two principal methods of valuation.

Where land has a potential for development, it is valued by the residual method which assumes the development to be completed and let and values it as such and then deducts all the costs, physical and financial, of achieving the development. The result indicates a ceiling for the amount which may be paid for the site.

Certain types of business depend crucially on location, for example, a petrol filling station. It is obvious that the throughput of petrol sales is the basis of the profitability of the operation, and the method of assessment is known as the profits method. Finally, where there is a situation where properties are not normally bought and sold and where the profit motive may not be the driving force of the work carried on in the building — a hospital or a town hall are two examples — a valuation may be approached by adopting a suitably depreciated estimated building cost and adding the land value (restricted to the current use). This final method is known as the contractor's method or basis.

Wherever possible, and particularly in the case of the residual, profits and contractor's methods, the result will be checked by use of one or both of the other methods, where practicable.

The remainder of this book is concerned with detailed considerations of the methodology and operation of each method in turn.

The comparative method

PART TWO

The market

Where a market exists which is both active and well publicized (a perfect market) it produces a pattern of prices which is informative to observers of the market and persuasive to would-be participants in that market.

In order for there to be a market there must be both buyers and sellers. The fact that there are shades of opinion is what constitutes a market. If everyone in the market held the same views, there would be minimal activity and consequently little evidence of the market. As it is, the prospective purchaser is able to observe the market; he can discover the price level and price range on any particular trading day and also the number and size of bargains struck at that level. He can monitor price movements over a period, compare performance across the market, and read financial columnists in their daily columns. In fact, he has access to what approaches a perfect market.

One such market is that in shares. Large companies have millions of shares and thousands — perhaps tens of thousands — of shareholders. Some of the holders of the shares may decide to sell all or some of their holding on a particular day. Their motivation is varied: the seller may need the liquidity offered by a sale; he may feel that the current price is higher than is justified by the performance of the company and that he should take his profit before the level falls; he may be dissatisfied with the return or sell his shares simply to transfer that cash into the shares of some other company which he judges to be more attractive.

The fact that buyers exist shows that not everyone reacts in the same way. Whilst the seller may expect the price to fall, the buyer is presumably optimistic of future performance based on a purchase at the current level. Whilst the seller believes that he can employ his cash better in other shares, the buyer anticipates that he will obtain a superior return by transferring some of his money to the shares under consideration.

For any goods item in common use, whether durable or consumer, production of identical items is likely to be numbered in hundreds, thousands or even millions. If the price of a particular brand of coffee increases, the consumer will look at other brands; where their prices have not changed, he may decide to switch to another brand. At some stage general inertia, brand preference and brand loyalty will be breached. Where the increase is due to a rise in the price of raw materials common to all brands, the consumer may consider changing to an alternative, say tea. Similar considerations apply to all other economic goods such as butter, newspapers, refrigerators, clothing, cars, etc.

An important though not necessarily the only consideration in making a choice will be price. The evidence of the interaction of the open market is likely to provide the best evidence. Producers pricing their product much higher than a similar competitor's product run the risk of losing turnover. Where one share is considered too expensive by a potential buyer, he will switch to other shares.

This is also the case with property. Where there is a street of similar houses, none of which has any additions or alterations and of which six have all sold within a few months at similar prices, there could hardly be better evidence of market value. Similarly, where an office block, warehouse or shop is to be valued, information on recent transactions involving similar nearby properties will be of considerable assistance.

In reality the property market is much more complex than that. In the first place, no property is identical, if only because each building occupies a different site. Even where two buildings were designed and constructed in identical fashion, one may have been improved or abused in a way in which the other has not. It may face in a different direction, enjoy a better outlook or be located on the opposite side of the road. In some cases, what appear to be minor differences may assume critical importance. In particular, the precise location of a retail shop is of crucial concern. An intervening junction, a pedestrian crossing, a site on the opposite side of the road or even a difference of 20 or 30 m may well influence the view of prospective retailers as to the desirability of the premises and therefore affect the rental and capital values.

Not all the differences are physical or visible. Two properties of identical appearance may well differ in tenure, one being a freehold interest and the other leasehold; depending on the details, the latter is almost certain to be worth less, and in some cases substantially less. Information is needed on the existence of onerous covenants in the lease (or for that matter in the freehold title, though these are less likely). A restrictive planning permission is likely to have an adverse effect on value, whilst the existence of planning permission, particularly for such 'bad neighbour' uses as a scrap yard, a fish and chip shop or an amusement arcade may both enhance the value of that land and have a depreciatory effect on the value of adjoining premises.

These introductory remarks serve to suggest that the property market is unlike any other volume market and needs its own treatment. The valuer must be aware of returns available from non-property investments which will vie with property for the attention of investors. On the one hand he is providing the information on which comparisons with other investment media may be made; on the other he is attempting to distil from information about property transactions those elements which will be useful in building up and assigning value (both rental and capital) to other premises.

Finally, it is worth emphasizing that the professional investors accounting for the major part of market activity are seeking the maximum return possible. They will be prepared to move into or out of particular investment in stocks and shares according to their view of the future of the market. They may decide to remain liquid; that is to place cash holdings on 7-day deposit at the ruling rate of interest. Any sizeable investment fund will have views on the spread of investments and, if large enough to own property, the maximum amount to be invested in property. To this extent therefore property is being measured not only against similar property transactions but against the returns on other investments.

5 | Approach to comparison

Capital value is determined by a number of factors: the present and prospective income; the return which the market determines to be appropriate to the particular investment; the strength of the tenant's covenant; and the lease terms and the tenure of the property. It is therefore appropriate that in seeking to analyse real property transactions involving similar properties to the one to be valued, particular attention should be paid to these determinants. They are independent of each other but at the same time intricately interwoven; interpretation is often extremely difficult, involving aspects of judgement and experience. The determinants referred to above will now be considered. A more complete discussion will be undertaken in Chapter 8.

5.1 DETERMINANTS OF VALUE

5.1.1 Income

Where a recent transaction is being analysed, the current rent payable will be known. It must be determined whether the rent is the same as or lower than the rental value and the extent to which the lease terms erode the income.

5.1.2 Duration

Most investment properties (with the exception of residential properties) are usually let for substantial periods. In the past, the length of letting was a major indication of stability in real estate: investors preferred a term of 21 years to one of 15 years (other things being equal) and would be delighted to conclude a lease for a term of 30 years. Until the late 1950s such terms were often granted at a rent fixed for the whole of the period or with modest increases determined at the commencement of the lease.

With the onset of significant inflation, the rents in such lettings began to be less attractive to investors and to have an adverse effect on capital values.

For the landlord–tenant relationship to work satisfactorily, very short terms are impracticable. The lease of a substantial building typically requires the tenant to maintain the fabric of the building, including painting at stated intervals and repairing and renewing where necessary the fabric or parts of it and the services such as electricity installations and water supplies, lifts and boilers (either directly or through contribution to a sinking or reserve fund).

Unless the tenant is committed to occupy for a fairly long period the maintenance cycles become more difficult for the landlord to impose and enforce and more onerous for the tenant who eventually carries out the work. An expensive lift renewal, for example, would make the building completely uneconomic to a tenant if reflected over his occupation for a short period such as 1 year whereas it is much more acceptable and reasonable if seen in the context of a lease for 20 years or more.

From the tenant's point of view, a long lease is equally important. It is likely that when he takes the premises they are basically suitable for his purpose but need some adaptation or alteration if his business is to be run with maximum efficiency and security. For example, a solicitor looking for practice premises will be interested primarily in their location; are they in a 'professional' area near other solicitors and the courts and also in the area of other professionals such as accountants, surveyors and estate agents? He will resist any property which does not pass this test but is unlikely to refuse an otherwise suitable building because it does not have a strongroom, even though it is an essential feature of his occupation. The construction of a strongroom may cost several thousands of pounds but he will be willing to undertake the work in return for a suitable length of lease over which to 'write-off' the expenditure. Although he may be lucky to find suitable premises with a strongroom, to expect to do so would limit his search greatly. Similarly a butcher is likely to be faced with providing a cold store, and a jeweller security grilles and shutters.

Tenants will wish for a longer term for a more fundamental reason — the retention of goodwill. A tenant will be reluctant to move his business once it has become established in certain premises as to do so will risk losing some of his customers.

It will be seen that both landlords and tenants have good reason for regarding the landlord–tenant relationship as a long standing one. So the dilemma of depressed rental and capital values facing owners and the consequent reluctance to grant long leases was resolved by the introduction of rent reviews. The lessor grants a fairly long term at an initial rent but with provision for a reconsideration of the rent payable at intervals in the light of general rent movements. The institutional lease is normally for a term of 25 years with reviews at 5-yearly intervals. The parties may of course agree whatever terms they wish and 7-year intervals are quite common. Where particular properties are in great demand or where the market is demand-led (as in the early 1970s) the landlord may find it possible to require a 3-year interval. Most review clauses specify that the rent will not decrease — the effect of a standstill or downturn in rental values is that the tenant continues to pay the same amount even where the rental value has fallen below the earlier level. By this means the landlord has a valuable assurance of a minimum level of return from his investment whilst the lease continues, unlike an investor in a company who may find a dividend passed after a difficult trading period.

The effect of rental growth and the rent interval is that both parties benefit to some extent. At each review, the landlord achieves a new rent based on current rental values whilst the tenant is assured of a fixed rent until the next review. Where rental values continue to grow, the tenant has a valuable interest in the property based on the fact that for the greater part of each review interval, he is paying a rent less than the market rent.

5.2 EXPIRY OF LEASE

At the end of a lease, the tenant is free to leave but if he decides that he wishes to remain a good deal of protection is afforded by the Landlord and Tenant Act 1954 as described in a later chapter. As was discussed above, the rent paid for most of the time during the currency of a lease is below rental value. One of the considerations in analysing the components of capital value is the rental value where this is not the same as the rent payable. The investor and the valuer will have a view as to the rental value although aware that it cannot be charged until the next rent review or where no reviews are scheduled, the expiry of the lease. Some of the evidence examined will relate solely to rental value. Recent transactions in respect of comparable properties will be analysed in an endeavour to find an appropriate rental level for the property under consideration.

5.3 ASSESSMENT OF RENTAL VALUE

Several difficulties arise including the common one of lack of information. Rarely is there sufficient information for a statistically significant sample to be taken. Much of the information available will be no more than indicative. In other words, the premises may not be wholly comparable in terms of, for example, size, location, age, or standard of construction.

The lease terms are unlikely to be precisely similar. The valuer is accustomed to such problems and his approach is that any information is better than none. The degree of reliance placed on any information available will vary according to its nature.

The circumstances in which the comparable rent was achieved are also of some relevance. A rent for a new letting in the open market should be a good indication of the market but the valuer will need to consider the circumstances leading to the agreement. There is a tendency to agree rent-free periods. This is sometimes done to reflect the often high cost of fitting-out premises to bring them to a state suitable for occupation. This may be the case where the landlord offers a shell only, with internal finishes and services to be supplied by the tenant or where, in existing premises, the tenant undertakes to remedy defects or bring the accommodation up to a satisfactory standard. A landlord will prefer a rent-free period to the alternative of a rent reduction which would not only operate throughout the whole period to the next review but would reflect on the capital valuation of the premises. Where it is possible to negotiate, the tenant will look for a significant acknowledgement of his cost and a 5 or 10% reduction in rent (albeit for the whole of the period to the first rent review) is likely to prove unattractive against a totally rent-free period of 3 or 6 months.

5.4 RENTAL EVIDENCE

Rental evidence derived from lease renewal or rent review negotiations should be examined closely. At the end of a lease a business tenant is normally entitled to a renewal (s.30 of the 1954 Act sets out the only reasons available to the landlord to make a valid objection to the grant of a new tenancy where renewal is requested by the tenant). The rent is fixed by agreement between the parties but where they are unable to agree there is provision for the courts to determine the terms, including the new rent (the latter under s.34).

This benefit of interpretation is available to the tenant on renewal under the Act and is likely to result in a rent somewhat less than might have been achieved in the open market. The Act instructs the court to ignore tenants improvements in certain circumstances.

5.5 RESTRICTIONS ON USE

The court is also likely to interpret a restrictive user clause or other provision in the original lease as a disadvantage to the tenant, and therefore to be reflected in the rent fixed, so careful enquiries need to be made before accepting a rent determined under section 34 as evidence of the market rent.

Similarly, on review (which is carried out not under the Act but according to the provisions of the lease) the rent fixed may depart from market value as agreed in the lease. Provisions for settlement of disputes over the level of the new rent are in the hands of an arbitrator or expert surveyor as laid down in the lease.

In either case the task is to reflect the market rather than to make it and the result of imperfections in the information gathering process is often to soften to some extent the level of rent fixed.

5.6 LEASE TERMS

The lease terms are important on two counts. First, they determine the extent to which the rent paid is the net rent received by the landlord. Where, under the terms of the lease, he is responsible for certain repairs or for the payment of insurance premiums, the rent paid is not a net rent and the property will require more management input (itself an expense to the owner). Second, institutions normally insist on excluding from their portfolios any properties where the tenant is not subject to a 'full repairing and insuring lease' and therefore responsible for all outgoings. So a perfectly sound investment occupied by good tenants may not find a market amongst the main purchasers of good quality investment property because of an inappropriate lease. The effect of a loss of a significant sector of the potential market is of course likely to be reflected in the capital value of the property.

There is also the important point that net income is eroded progressively from one review to the next by the effect of inflation. As an example, assume that the landlord's cost of repairs and insurance is assessed as being 10% of the rent agreed. Further assume that the increase in costs of such outgoings is 10% p.a. Although the rent is fixed for the review period, the cost of outgoings is not and there is evidence that the increase in the cost of maintenance work is greater than that for new work especially when the economy is buoyant and there is considerable building activity. Under these cirucmstances the net income over a 5-year period will decrease as a proportion of rent payable as shown in Table 5.1.

5.7 THE TENANT'S 'COVENANT'

In any lease, it is important that the tenant is of sound financial standing. A 'blue chip' public limited company will be much sought after as a

Table 5.1 Effect of inflation on net income where landlord meets cost of some or all of the outgoings

End of year	Rent payable (£)	Cost of outgoings as % of rent including inflation	Net income as % of rent payable
1	100	10	90
2	100	11	89
3	100	12.1	87.9
4	100	13.31	86.69
5	100	14.64	84.36

tenant of premises because of the undoubted ability and readiness of the company to comply with the lease terms, notably by paying the rent promptly and maintaining the premises to a high standard. The standing of the tenant is particularly important when there is an economic down-turn resulting in difficult trading conditions and the possibility of trading losses. A small company or a company experiencing rapid growth could find itself sufficiently short of funds to put the payment of rent at risk. A well-established company without high borrowings would 'weather the storm' by drawing on reserves.

5.8 ANCHOR TENANTS

A quality tenant is highly valued particularly when the retailer agrees to take space in a new development. The knowledge that one or more well-known traders has agreed to be represented in the new development acts as a seal of approval. It gives confidence to the smaller retailer that there will be a level of activity in the development generated by the presence of the larger retailing group or groups. Aware of this power, the retailers may well negotiate a lower rent or other allowances or inducements for the initial period of occupation.

5.9 THE YIELD

The yield required by an investor on his investment is always a complicated process and tends to be even more so in property. The basic analysis is straightforward, at least in the case of freehold interests let at full rental value. Where the investor in a freehold property expects a return of say 8% on his investment, the net rent represents that return; in other words, the rent is 8% of the capital value. The capital value itself is found by dividing the rent by 0.08 (the decimal fraction indicating 8%). Alternatively the same result is obtained by dividing 1 by 0.08 to find the reciprocal (otherwise known as the 'years' purchase') being the multiplier needed to convert annual income to capital value. In this case the years' purchase is 12.5 (1/0.08) and the capital value of a net rent of £1000 p.a. is £12 500.

The analysis of a transaction is largely the reverse of a valuation and consists of decomposing the sale price achieved.

Published results of property transactions show average all risks yields from 5% upwards and even lower initial yields; this immediately raises the question as to who would or should wish to invest in property to show such a low return when the same amount of capital could be invested safely in government stock or bonds, banks accounts or building society investments and relatively safely in shares in major publicly quoted trading companies to show much higher returns.

The investor would need to understand the 'shorthand' employed here. Buying a freehold property let at a rack rent to show an 8% yield is only part of the story. One of the reasons for purchasing a property investment is the anticipation that the rent will increase at least in line with inflation and that such increases may also increase the capital value. It is therefore apparent that the willingness to accept an 8% return is heavily qualified; by projecting rental value it can be shown that the anticipated return is significantly higher than 8% provided that the lease is nearing its end or incorporates regular rent reviews, either of which will allow the rent to be renegotiated.

5.9.1 The all risks yield

The property return is misleading since it follows conventional valuation practice of assuming that the rent received will not at any time exceed the current rental value, although the expectation of growth is the most powerful reason for investment in property. The unexposed growth is then compensated by adoption of a low all risks yield. The alternative approach of anticipating rental growth and discounting receipts at the higher (real) market rate of interest is possible but has so far not won much support from the profession. Analysis in which the growth expectations are made explicit is discussed later.

5.10 ANALYSIS

There are two main themes to analysis: yield and rental value. To determine a yield it is necessary to know the rent paid, the rental value (where there is a difference), details of tenure and the price paid. Where this information is available, analysis of an investment transaction will indicate the purchaser's requirement for a particular level of return. Where the transaction has taken place in the open market it may be assumed that the result is an indication of the market reaction to the advantages and disadvantages of successive streams of income issuing from that type of property.

The process of analysis is a reversal of the process of building up a valuation which has been described earlier. It is a fundamental requirement that the approach to analysis and valuation should follow the same rules. Any departures will question the internal consistency of conclusions drawn from particular transactions and applied in valuations of other properties.

Example 5.1

Office premises were recently let on completion of building works at a rent of £70 000 on FR and I terms to a well-known public company and the freehold interest immediately sold to an investment company for £1 000 000.

Commentary
The information given enables the valuer to carry out an analysis to deduce the initial yield which is also the all risks yield in this case, the latter being the yield which implicitly takes account of future changes in rental income.

The yield may be found by $R/C \times 100$
where R = rental value
C = capital value or sale price
In the example $R/C \times 100 = 70\,000/1\,000\,000 \times 100 = 7\%$

The tentative conclusion is that the purchaser has achieved a return on capital invested of 7%. The basis of the analysis is that the result of this calculation may be used in assessing the capital value of similar premises.

The next stage is to explore why a 7% yield was achieved. How much was contributed by the development and to what extent was it influenced by the quality of tenant attracted or by the terms of the lease?

Example 5.2

A similar building let to a substantial private company having a business operating in the region and let on similar terms was sold for £875 000.

Commentary
Undertaking a similar analysis, the yield can be shown to be 8%. The conclusion must be that the market is prepared to purchase the investment provided that the small additional risk represented by the less

substantial nature of the company is recognized and compensated for by an increase in yield.

Not only are these examples at the simplest level of the problem encountered, their treatment is also too simple. The investor wishes to know his real return and so would take into account any periodic out-goings and any initial costs associated with the purchase. It is therefore necessary to know the contents of the lease and the liability for expenses relating to the building. It is usual to let modern commercial premises subject to the tenant being responsible for all repairs and maintenance and for insuring the building against fire or other damage and loss. The investor will be left with the cost of management for which an agent will usually be employed. His work will include collection of rent, periodic inspections to ensure that the tenant is observing the covenants of the lease, liaison to ensure that the building is insured at a proper value and the negotiation of a revised rent on review or when the lease is renewed at the end of a lease. Rent negotiations may or may not be included in the management fee agreed between the parties but it is important for the estimated annual cost to include a sum for this work also. For the purposes of valuation and analysis, the cost is usually taken as a per-centage of the rent, the rate depending on the type, etc. The initial costs of purchase will include stamp duty, legal charges and disbursements and surveyors' charges for advising on the purchase price and possibly nego-tiating the purchase. Precise costs will depend on all the circumstances but are usually taken as 2.75–3% of the purchase price. When this information is introduced into the analysis the yield is affected.

Example 5.3

Take the information from Example 5.1 where the total cost of purchase including fees is £1 027 500.

The annual management charge is 2% on rent or £1400.

Commentary

	£
Rent received	70 000
less management	1400
Net rent	68 600

$$\frac{68\,600}{1\,027\,500} \times 100 = 6.67\%$$

This shows the real return to be one-third of a percentage point lower than calculated earlier.

It is clear from these examples that the inclusion of initial and continuing costs affect the yield. Analysis taking account of these items is clearly preferable and is essential if a true return is to be calculated.

It is important that both analysis and valuation based on information dervied from that analysis should adopt the same assumptions.

Example 5.4

A smaller office block built 25 years ago is let at £24 000, the landlord being liable for external repairs and insurance. The rent was fixed following a review 3 months ago. The tenant is a local company. The freehold interest has been sold for £220 000.

Commentary
The rent needs to be adjusted to reflect not only the cost of management but the repairing and insuring responsibilities of the landlord. The net rent is

		£
Rent reserved		24 000
less management 4%	960	
repairs @ 5% of rental value	1200	
insurance @ 2% of rental value	480	2640
Net rent		21 360

$$\text{Then } \frac{R}{C} \times 100 = \frac{21\,360}{226\,050} \times 100 = 9.9\%$$

Factors likely to have influenced the higher yield achieved include the status of the tenant company, the greater management involvement and the liability for outgoings. In times of inflation, the annual costs of repair and insurance are likely to increase, eroding the net income of the landlord.

5.11 EVIDENCE

Recently agreed rents are a good source of information. The process of analysis varies according to the type of property concerned. In analysing the rents of office buildings, the valuer will take into account the number of floors in the building and facilities available such as lifts, air-conditioning, raised floors and car parking. It is possible that the lower floor levels will command better rents than the higher floors but whether this is so and whether the differentials are appropriate is a matter for careful analysis and judgement in each case. Knowledge of the rental value of a particular floor is not helpful unless the floor area is known, as size is likely to have an effect on unit values. The almost universal practice is to quote rental values in terms of rents per square foot (or square metre).

Information on office rents is widely quoted but the valuer should be careful to enquire the basis on which the floor area has been computed. The area may include all the space on a particular floor, the gross space within the unit, the net space after allowing for walls and partitions or the areas occupied by rooms only. In well designed offices, it is expected that the net usable area is 90–95% of the total area but in poorly designed blocks or where the building has been converted from another use, the wastage may be far higher.

Shop rental values are also reported on a unit basis. It is common for the sales area to be weighted to reflect the view that a greater volume of trade will be done in the part of the shop nearest the street, and the initial depth will have a higher rental value. There is a convention that sales areas should be zoned in 20′ or 6 m zones parallel to the street with decreasing rent levels applied to each zone. In the most used application, the first zone, zone A, is assigned the highest unit value, the next zone, zone B, is taken arbitrarily at half the unit value of zone A and the remainder, zone C, is taken at half the value of zone B. This is termed 'halving back'. There are variations involving different depths of zones and additional zones but the principle is the same. It is important that analysis and application to other shops is undertaken on the same basis. The result is a value per square foot or square metre in terms of zone A (ITZA). The approach is widely recognized and used by valuers and has been accepted by the Lands Tribunal on countless occasions especially in rating valuation cases. It is only in recent times that the veracity of the approach has been questioned which is surprising considering its arbitrary nature. The application will be described by means of some examples.

In the period before shopping centre developments became common, many developers, tenants and surveyors recognized a standard shop unit as one having a frontage of 20′ and a built depth of 60′. The depth lent

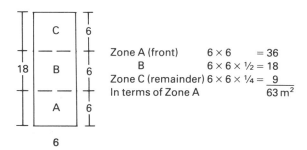

Figure 5.1 Overall floor area of 108 square metres reduced to equivalent units.

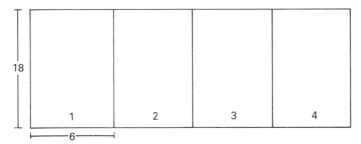

Figure 5.2 Four identical 'standard' shop units.

itself to division into three zones of equal (20′) depth giving an overall floor area of 1200 units and an area in terms of zone A of 700 units. The smaller unit value reflects the tapering off of values from front to back as shown in Figure 5.1.

Assume that the shop unit is let at a net rent of £8400 p.a. Then the overall unit value, based on an area of 1200 ft^2, is £7 per ft^2. On a weighted basis, the value is £12 per ft^2 for zone A, £6 for zone B and £3 for the remainder. The equivalent weighted figures per square are £133.33, £66.67 and £33.33. The figures and the following examples are worked in metric units although imperial measurements are still commonly used.

Example 5.5

Consider four identical 'standard' shop units, all available to let at £8400 each p.a. as shown in Figure 5.2.

Prospective tenants include a butcher, a jeweller, a newsagent and a dry cleaner. The butcher requires more space to accommodate a cold store and cutting room whilst the jeweller needs a maximum of 12 m depth. The dry cleaner would be satisfied with half the frontage and has introduced a shoe repairer who wishes to operate a small instant heel bar. The newsagent would find another 6 m in depth very useful and land is available at the rear for extension. These requirements are shown in Figure 5.3.

A comparison of the rents charged depending on whether an overall unit rate or a zoned rate is charged is shown in Table 5.2.

If the reasoning associated with zoning is correct, the effect of adopting an average unit rent would be to undercharge those tenants with less than standard depth (jeweller) and overcharge those with additional area towards the rear (butcher, newsagent). Furthermore, it is accepted that a small kiosk type unit is likely to command a higher unit rate than represented by zone A figures since it can concentrate its activity in a very small area but still benefit from exposure to shoppers using the street. From the analysis, it appears that the landlord would let the jeweller's

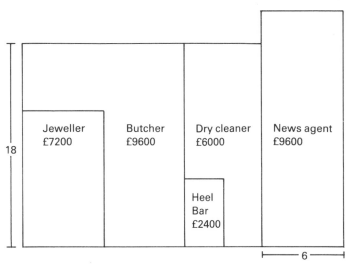

Figure 5.3 Rental values by zoning.

Table 5.2   Effect of overall rate

Trade	Frontage (m)	Depth (m)	Total area (m²)	Area ITZA (m²)	Rent @ £13.33 Zone A (£)	Rent @ £77.78 overall (£)
Jeweller	6	12	72	54	7 200	5 600
Butcher	6	18	144	72	9 600	11 200
Dry cleaner	3	18	90	45	6 000	7 000
Heel bar	3	6	18	18	2 400	1 400
Newsagent	6	24	144	72	9 600	11 200
Total					34 800	36 400

shop very quickly at £5600 but could not make up the balance of the standard unit rent of £2800 by letting space to the butcher assessed on an overall rate basis since, on the zoning basis, the rear area is worth only £1200 extra whereas as part of the jeweller's shop £2800 has been deducted. Other factors are present: the dry cleaner might expect a lower rent than calculated because he will have a narrow frontage and a relatively large amount of rear space.

More recently it has been argued that the method has a weakness since the zone depths are fixed arbitrarily by the valuer. It this argument is pursued it will be seen that the valuer can claim to have market evidence only where he analyses transactions using information from actual depths, termed 'natural zoning'. The information may be used by limiting the zone depths to the actual depth for which direct rental evidence is avail-

Table 5.3 Effect of natural zoning

Zone	Trade	Frontage	Depth (m)	Area (m²)	Rent (£)	Per square metre
				Area beyond natural zone		
A	Heel bar	3	6	18	2400	133.33
B	Jeweller	6	12	36	7200	100.00
C	Dry cleaner	6	18	18	6000	66.67
C	Newsagent	6	24	36	9600	66.67
C	Butcher	6	18	36	9600	66.67

able. The effect of natural zoning for the example considered is shown in Table 5.3.

Zoning should be seen as an aid to analysis and its limitations recognized. The number of zones is normally restricted to three; zones A and B and a remainder. Most traders will view increasing depth as affected by the law of diminishing returns; there comes a point where they will offer no more rent for any additional depth and an existing shop with appreciably more depth than demanded may be difficult to let. From the investor's point of view, it would be foolish to extend the depth of the shop unless an adequate additional return on the cost could be anticipated.

It has been suggested that the analysis of retail rents should pay less attention to zones, whether arbitrary or natural, concentrating on the relationship between frontage and depth. An analysis of overall unit rental values should be carried out. It would be expected that the greater the depth, the lower the overall unit rent.

If the likelihood of a relationship between frontage and depth is accepted, it would be possible to make predictions, using the theory of regression analysis. Given enough information, which is always a problem in valuation, the line minimizing the sum of the squared deviations from that line can be found and is referred to as the line of 'best fit'. The line should give useful indicators for the relationship between frontage and depth and the effect on rental value.

Other approaches to analysis include other units of value such as a hotel room or a school place (either may be analysed in terms of rental or capital value, depending on the type of transaction information available and the purpose of the valuation). Valuations of this nature are specialized and the valuer has a detailed knowledge of that segment of the market; the price per unit (bedroom, etc.) is a rule of thumb which results from extensive experience of the market.

Dwelling houses are usually sold rather than let and most often they are purchased for occupation not investment. Valuation is normally undertaken by comparison of capital values but the valuer should recognize, even though unable to quantify, that preference and sentiment will play a part in the price paid. The sale of private dwellings is usually undertaken by someone who practises in a very localized area and who has a good general knowledge of the market activity in that area. The valuer will know the preferred estates and the favoured roads in that estate, the demand for different sizes of dwelling and the level of amenity usually demanded. The addition of another bedroom to a four-bedroomed house may not increase the value very much whereas a three-bedroomed house given an extra bedroom may have a substantial effect on value; the installation of a modern central heating system in a small bungalow may prove to be a sensible expenditure in terms of obtaining a satisfactory price for the property and making it easier to sell.

Example 5.6

A post-war residential development is well sited 3 miles to the south-west of the city centre and there is a frequent bus service to the centre. The majority of the houses are three-bedroomed semi-detached with central heating and a single garage. Many of the owners have undertaken improvements to their houses. The price for an average unimproved house is approximately £120 000. You have been instructed to value a house recently placed on the market. It is a typical house for the estate except that it stands on a larger corner plot giving space for a second garage. The original attached garage was demolished and a double garage and laundry constructed in its place with a double bedroom above part of it. At the same time an *ensuite* bathroom was added to the main bedroom. Office records disclose numerous transactions, including the following in the last 6 months.

Address	Specification	Price (£)	Comments
10 Allens Way	3 bed S/D CH S/G	120 000	
15 Allens Way	3 bed S/D CH S/G	115 000	Quick sale required
18 Allens Way	3 bed S/D CH S/G	123 000	Backing on to golf course, large garden laid out professionally
25 Allens Way	3 bed S/D CH D/G	125 000	
27 Allens Way	3 bed S/D CH D/G	124 500	
29 Allens Way	3 bed S/D CH S/G garden room	121 000	
12 Bens Lane	3 bed S/D CH D/G laundry	126 500	
18 Bens Lane	4 bed S/D CH D/G	134 000	
13 Bens Lane	4 bed S/D CH S/G	128 000	
17 Bens Lane	4 bed S/D CH D/G	150 000	One of 10 built on the estate

Commentary

The basic house price is confirmed at £120 000 by the information given. An additional garage is seen to attract an extra £4500–5000, whereas a garden room or laundry seems not to be required (the additional value of £1000–1500 is well below even the most modest cost of construction). A fourth bedroom appears to add between £8000 and £9000. The price of the four-bedroomed detached house (presumably designed as such) suggests that the demand for four bedrooms is associated with the need for more space on the ground floor and possibly a larger site.

On the basis of the information available, the valuer could justify a valuation of £134 000–135 000 (including the *ensuite* bathroom).

Similar considerations apply to business premises.

Example 5.7

You are required to give your opinion as to the capital value of a 1950s factory of $1000\,m^2$ and offices of $100\,m^2$ currently vacant and to which your client wishes to transfer his thriving engineering business.

The following information is available from office records and other sources.

Factory A: Area of $2000\,m^2$ including $150\,m^2$ of office. Modern building recently let at £55 000 p.a. with 25 FR and I lease, and 5-yearly reviews.

Factory B: Area of $2500\,m^2$ plus $300\,m^2$ of office accommodation above part. Built in 1970s. Let on modern type lease and rent recently reviewed to £65 000 p.a. for next 5 years.

Factory C: Area of $750\,m^2$ without office accommodation. Pre-war accommodation on cramped site let @ £11 250.

Factory D: Recently completed basic factory units of $1500\,m^2$ letting readily at £28 per m^2. The developers are prepared to provide the area of office space required by the individual tenant at £60 per m^2. The units may be purchased to show 7.5% yield but none has so far been sold.

Factory E: An old factory with a floor area of $3000\,m^2$ in good structural condition has recently been let @ £16 per m^2 and sold immediately afterwards to a local investor for £345 000.

Commentary

Yields range between 7.5% and 17% whilst rents of factory space only range £16–34 per m^2. Office space is required by most businesses but probably more than 10% of the factory floor space is excessive. The more substantial construction and superior finish and fitting of offices would justify a higher rent than that obtained for the factory floor space. The adjustment to be made is a matter of judgement in each individual case.

In this analysis, it has been accepted that the appropriate relationship between factory space and office accommodation is 10:1 and that the office rental value is 50% above that of the associated factory space.

The analysis provides the following information.

1. Factory D (new) suggests a yield of 7.5% (although no sales have taken place).
2. Factory E (old but sound) — the sale shows a yield of 14%.
3. The rental of Factory D shows a ready market for new units @ £28 per m^2. An older and smaller unit with a poor site has a rental value of £15 per m^2 (Factory C).
4. Factory A suggests a factory space rental value of £15 per m^2 whilst Factory B suggests a slightly lower figure which could be attributable either to the increased size or to the fact that it is older than A and less attractive. Current development of units of 1500 m^2 provides some evidence that the main demand is at the lower end of the scale.

The various conclusions enable the valuation to be made as follows.

		£
Factory space	1000 m^2 @ £19 per m^2	19 000
Office space	100 m^2 @ £25 per m^2	2500
Rental value		21 500
YP perpetuity at 12%		8.33
Capital value		179 168
	say	180 000

Commentary
Given more information such as lease terms, quality of building, site area and demand, it would be possible to tabulate the information for ease of comparison.

It will be seen that the resulting valuation is more than a mere mathematical calculation. The valuer has used his judgement in applying information available on recent transactions to deduce a rental value of £19 per m^2 of factory space and £25 per m^2 of office space. He has then inferred a yield of 12%, again based on the information at his disposal. It is unlikely that another valuer would arrive at precisely the same set of conclusions but the resulting capital values could be expected to be similar.

Where the reversion is delayed for a longer period than a normal rent review interval, the use of the all risks yield is unsatisfactory since the assumption as to regular growth do not apply to the current rent, which is fixed for a relatively long period.

Example 5.8

A shop unit is currently let at £25 000 p.a. on a lease without review having 12 years unexpired. The current rental value is £45 000 p.a. and it is estimated that rental growth of 6% p.a. will continue indefinitely. The freehold was sold recently at a price of £650 000. The appropriate yield for similar rack rented units on modern lease terms including 5-year upward only reviews is 5%.

Commentary
A unit let on modern lease terms at the current rental value would have a capital value of £900 000 made up as follows.

	£
Net rental value	45 000
YP perpetuity at 5%	20
Capital value	900 000

The value of the transaction quoted is approximately two-thirds of this figure for two reasons: the current rent is £25 000 p.a., much below the current rental value; and there is no opportunity to benefit from the increasing rental trend for the next 12 years because the lease does not provide for any review during this period.

An attempt to analyse the transaction may be made as follows.

		£
Purchase price		650 000
less value of reversionary interest		
(assumed to be on modern letting		
terms with regular reviews)		
current rental value p.a.	45 000	
YP perpetuity at 5% def. 12 years	11.14	501 154
Balance, being share of capital value		
attributable to value of fixed income		
for next 12 years		149 846

The yield on this part of the investment is approximately 12.5% and could be compared to the yield on undated gilts subject to an additional risk premium.

The investment method

PART THREE

The determinants of value | 6

6.1 INTRODUCTION

The purpose of an investment valuation is to provide an opinion as to the capital value of the right to receive annual streams of income. It may be prepared on behalf of the owner in the form of advice preparatory to marketing, for a prospective purchaser or for a third party contemplating the grant of a loan secured on the property.

The majority of valuations are prepared on the basis of market value but the valuer should ascertain the purpose of any valuation as it may have an effect on the result or more particularly on the way in which it is reported. The valuer may be instructed to report on the worth to the owner or on some other 'non-market' basis: in such cases the report should make clear that the valuation has been prepared having regard to particular instructions and that the result reported does not necessarily represent the market value of the premises.

The consideration of capital value involves the collection of many details and the formation of a view as to the quality of the investment as a prelude to the necessary mathematical calculations to produce a valuation. The former part of the process relies on the skill and judgement of the valuer, whilst the latter is a relatively straightforward processing of figures based on financial formulae; the main formulae are listed in Figure 6.1. The mathematical calculations are readily adaptable to execution by computer program although the valuer should ensure that the formulae and valuation models are correct and appropriate to the particular task in hand. It is also necessary to determine what is being valued. Each separate property is unique; even where it is indistinguishable in form from an adjoining property it occupies a different site, the location of which may be of great importance in the consideration of value. The real importance usually lies in the quality and quantity of the particular income, in the limitations of title and in the constraints imposed by law, for example, on the use of the premises and on the landlord–tenant relationship.

1. Amount of £1
   The amount to which a single deposit of £1 will grow in a given number of years and at a stated rate of interest

   $$(1 + i)^n$$

2. Present value of £1
   Today's value of the right to receive £1 at some future date, given the rate of interest to be used

   $$\frac{1}{(1 + i)^n}$$

3. Amount of £1 p.a.
   The amount to which annual deposits of £1 each will grow in a given number of years and at a stated rate of interest

   $$\frac{(1 + i)^n - 1}{i}$$

4. Present value of £1 p.a. (YP)
   Today's value of the right to receive a series of annual payments for a given number of years, discounted at a stated rate of interest

   $$\frac{1 - \dfrac{1}{(1 + i)^n}}{i}$$

5. Sinking fund to produce £1
   The annual payment needed to accumulate to £1 in a given number of years and at a stated rate of interest

   $$\frac{s}{(1 + s)^n - 1}$$

6. Annuity £1 will purchase
   The annual income that will be purchased for a non-returnable lump sum payment of £1, given the number of years and the rate of interest to be applied

   $$i + \frac{i}{(1 + i)^n - 1}$$

also

7. Present value of £1 p.a. dual rate

   $$\frac{1}{i + \dfrac{s}{(1 + s)^n - 1}}$$

and

8. Present value of £1 p.a. dual rate adjusted for tax

   $$\frac{1}{i + \left[\left(\dfrac{s}{(1 + s)^n - 1}\right)\left(\dfrac{100}{100 - t}\right)\right]}$$

9. Present value of £1 p.a. in perpetuity (YP perpetuity)

   $$\frac{1}{i}$$

10. Present value of £1 p.a. in perpetuity deferred n years

    $$\frac{1}{i (1 + i)^n}$$

where
i = interest rate   } decimal
s = sinking fund rate } fraction
t = tax rate in pence

Figure 6.1 Valuation formulae.

Most aspects are open to judgement and will inevitably attract varying levels of interpretation and significance from one valuer to another. Such judgements may overlap the quantitative information previously referred to: for example, an expensively constructed building is not necessarily the best building from a tenant's point of view. The collective view of the market will be what determines the level of rental value. Whilst the form and content of the lease is known, the interpretation of its effect may be open to discussion. Similarly, the standing of a particular company as a tenant will be a factor open to interpretation by different valuers.

Perhaps most important, opinions as to rental growth will vary. Growth is a function both of the longer term future of the building itself and of the state of the economy, a complex consideration where any opinion is likely to rely heavily on the evidence of past performance for want of better information.

Matters capable of being quantified include the size of the building, the rent payable, the extent of the tenure and the provisions of the lease including the liability for any outgoings for which the landlord is liable. But even here there is room for opinion; what is the current rental value? what is the likely annual cost of outgoings for which the landlord is responsible?

Investigations and enquiries to obtain the information required should be exhaustive since the soundness of the valuation will be determined by the thoroughness of the collection and interpretation of data.

The status of a property in the investment market is an amalgam of these considerations which rely heavily on the interpretational judgement of the valuer. Market value cannot be determined without a good deal of investigation, both of the facts surrounding the particular building and of the circumstances in which similar properties were sold.

In support of the weight to be given to professional experience, Megarry J. (as he then was) had this to say about the expression of opinion by a valuer appearing as an expert witness.

'In building up his opinions about values, he will no doubt have learned much from transactions in which he has himself been engaged, and of which he could give first-hand evidence. But he will also have learned much from many other sources, including much of which he could give no first-hand evidence. Textbooks, journals, reports of auctions and other dealings, and information obtained from his professional brethren and others, some related to particular transactions and some more general and indefinite, will all have contributed their share. Doubtless much, or most, of this will be accurate, though some will not; and even what is accurate so far as it goes may be incomplete, in that nothing may have been said of some special element which affects value. Nevertheless, the opinion that the expert expresses is none the worse because it is in part derived from the matters of which he could give no direct evidence. Even if some of the extraneous information which he acquires in this way is inaccurate or incomplete, the errors and omissions will often tend to cancel each other out; and the valuer, after all, is an expert in this field, so that the less reliable the knowledge that he has about the details of some reported transaction, the more his experience will tell him that he should be ready to make some discount from the weight that he gives it in contributing to his overall sense of values. Some aberrant transactions may stand so far out of line that he will give them little or no weight. No question of giving hearsay evidence arises in such cases; the witness states his opinion from his general experience.' (*English Exporters (London) Ltd* v. *Eldonwall Ltd.*)

This summary provides a succinct account of the way in which any competent and careful valuer will go about his business.

As indicated above, the quality of an investment is a subtle combination of a number of distinct features, the precise effect of each on the particular investment being a matter of mature judgement. Various aspects of investment in property are now discussed in some detail.

6.2 OUTGOINGS

Landlords will favour a lease on full repairing and insuring terms under which the tenant accepts responsibility for all costs of repair and maintenance and for the insurance of the building, and its plant and fixed machinery against fire and other perils.

Where the property is let on such terms, the valuer will be concerned only with the costs of management incurred by the landlord. In other cases, he will need to make an estimate of the likely annual costs of maintenance or other expenditure for which the landlord remains liable, together with an estimate of the cost of accrued repairs.

The costs of management should include not only any annual management fee but annualized amounts for negotiations of rent reviews and lease renewals and for other services such as periodic assessments for insurance purposes where such work is not included in the basic management fee and is not recoverable by the landlord.

Approximate estimates of outgoings are sometimes achieved by taking percentages of the rental value, based on experience of managing similar properties. Such an approach should be used with extreme caution and then only when the information is not available in a more reliable way. Information gained from records of the maintenance costs of the property to be valued or in their absence from the cost records of similar properties should be used in building up a cost profile. The valuer should avoid presumptions and in particular satisfy himself that any information used is based on a satisfactory level and quality of work; he should also bear in mind the need to reflect the yearly equivalent of items recurring at longer intervals, for example, external decorations.

Landlords regard the full repairing and insuring lease (where the tenant is responsible for all outgoings) as an important factor for two reasons. The first is that detailed management involved in identifying and organizing work necessary is expensive and erodes the rental income. The second reason is that increasing costs over a 5-year interval (or until the next review) reduce the real value of the fixed rent. Some investors reflect their preference by seeking a higher yield where the landlord retains substantial repairing obligations; others consider only those investments where the lease is on net terms.

Tenants would be well advised to consider the condition of the building carefully before taking responsibility for such onerous terms. From the landlord's point of view, he will wish to be satisfied that the tenant is financially able to meet the probable demands of the covenant.

6.3 YIELD

The yield required by the market may be analysed from recent transactions involving the sale of similar investments.

The question of comparability requires careful consideration. Unless there is similarity in the various determinants of value discussed above between the property to be valued and the evidence obtained from recent transactions, there is a potential for misleading conclusions.

The size of the property, its age, the quality of tenant, the form of the lease (in particular its provisions relating to repairing obligations and the interval between rent reviews) and the period before the next rent increase can be implemented, all reflect opportunities for differences. The conventional approach to valuation, soon to be discussed, reflects all these similarities and differences in a yield analysed from known transactions and adjusted in the discretion of the valuer. The yield is then used to calculate a year's purchase (similar in effect to a price/earnings ratio). In the case of a perpetual income assumed to be derived from a freehold interest, the year's purchase is the multiplier applied to a net income to show the capital value of the right to receive that income in perpetuity. The multiplier is obtained by dividing the decimal fraction of the required yield into unity. For example, the year's purchase to achieve a return of 5% is 20 (1/0.05) and 8% is 12.5 (1/0.08). Where the income is to be received for a shorter term, the year's purchase for the same return is correspondingly higher. The reason is that the income is regarded as wasting and in compensation the return is increased to provide not only the return on capital but the return of capital over the period. The longer the term, the smaller the amount required for the replacement of capital; valuers commonly use this self-evident fact to assume that an income receivable for 50 years or more may be regarded as being received in perpetuity — the difference is very small, although it will vary with the rate of interest demanded.

The formula for year's purchase for a finite term is $\dfrac{1 - \dfrac{1}{(1 + i)^n}}{i}$ and the following comparisons of terms and yields will show the influence of the unexpired term.

YP at 5%
in perpetuity $1/0.05 = 20$
75 years $= 19.49$
50 years $= 18.26$
20 years $= 12.46$
YP at 8%
in perpetuity $1/0.08 = 12.5$
75 years $= 12.46$
50 years $= 12.23$
20 years $= 9.82$

A feature of the property market is the low all risk yields acceptable to the investor. For example, at a time when he could expect to achieve a return of 9.8% on undated government stock, the average yield on shop premises is 6.4% (itself the highest for a decade when for most of that time the yield has been below 5%).

However, whilst the return on gilts reflects directly the compensation required by the investor for foregoing the use of his capital, the rental yield is only one aspect of the property investment return.

As the rent increases, so the investor may expect the capital value to rise also (unless there is a compensating increase in the yield demanded by investors as has occurred recently in the case of shop investments noted above).

Where the market's fixed interest requirement can be ascertained, it is possible to calculate the income growth inferred by analysing the all risk yield. Further consideration will be given to this aspect when growth explicit methods of valuation are considered.

Valuers readily adjust the all risk yield to reflect perceived differences between the evidence deduced by analysis of recent transactions and the property to be valued. Adjustment in this way is a fairly blunt instrument and the valuer should consider carefully the effect of such judgements. Some valuers advocate the building up of a yield by a summation method consisting of first determining the risk-free yield, then adding premia for risk, illiquidity, the burden of management and any other aspect considered to detract from the quality of the investment. This approach begs two questions; first, what are the criteria for determining the risk-free rate?; and second, where is the evidence to support the values accorded to the other factors?

6.4 RENT AND RENTAL VALUE

6.4.1 Rent

The valuer will seek information regarding the current rent. Not only will he need to know the yearly rent payable but the frequency with which it is paid, when it is due, whether it is payable in advance or in arrears and what sanctions are available when payment is late (modern leases often contain provisions for a fairly punitive rate of interest of 3 or 4% above bank base rate to be payable on overdue payments). He must then enquire as to the other terms of the lease to determine the extent to which the landlord remains liable for any of the outgoings. In the ideal case (from the investor's point of view) the tenant will be responsible for all costs of repair, maintenance and insurance. Often the income is subject to outgoings for which the owner remains liable in which case the landlord must accept that the costs are subject to inflation and liable to increase on an annual basis, even though the rent itself is fixed between reviews.

There will be regular charges for management where an agent is employed and further legal and other professional charges may be incurred where disputes take place or where the rent is reviewed or the property re-let, either to the existing tenant or to a new tenant.

To this extent, the rent will be eroded and the market is likely to reflect this disadvantage and uncertainty by calling for a higher yield than required on a full repairing lease.

Value Added Tax will normally be payable on the rent (and any services supplied); where it is, the owner will have a duty to account to H.M. Customs and Excise.

Where the rent being paid is less than the rental value, the valuer will need to analyse available evidence to enable him to assess the market rent obtainable.

6.4.2 Rental value

Any valuation must take account of net income. Where it is anticipated that the rent payable will alter, this should also be reflected.

Estimation of the current rental value involves a judgement of the market, and adds a further dimension to the process of valuation. It should be noted that in the conventional approach to valuation, no attempt is made to project the rental value to the time when it first becomes payable. The valuer is simply required to estimate the current rental value and then to take it into account in his valuation from the earliest time that it may be imposed.

Rental value is estimated by an analysis of whatever relevant information of rent levels is available to the valuer. He should have a good knowledge of the market but may not always be aware of all the transactions which have taken place. Where he can establish a prevailing pattern of rental values, he will then wish to use it to reach a view of the rental value of the particular unit under consideration bearing in mind that not all the evidence may be relevant or relate to truly comparable properties. For example, if the unit is appreciably larger or smaller than the properties from which the basic information is derived, the valuer's judgement may suggest that there is a need for some adjustment. Age or other physical attributes may affect his view and he will need to consider the legal framework within which the property is let. If the current tenant is likely to seek a new lease he must take into account the effect of the Landlord and Tenant legislation and in particular exclude the value of any relevant improvements made by the tenant. The likelihood or otherwise of any unsatisfactory lease terms being changed in negotiations between the parties or by the court must be taken into account, including the effect on rent of any unusual review pattern or onerous repairing obligation likely to be replicated in a new lease.

6.5 THE LEASE TERMS

The lease is the contract between landlord and tenant that controls their relationship throughout its term and in some cases beyond it. A badly drawn lease is likely to be all too apparent in the case of any disagreement. The various provisions of a satisfactory lease will exert considerable influence over rental and capital values and are worthy of careful investigation.

Landlords aspire to a modern institutional type lease which may be described as a lease for a term of 25 years with provision for upward only rent review at intervals not exceeding 5 years and with the tenant responsible for the execution of all repair and maintenance work and for the cost of insuring the premises including cover for loss of rent in the event of damage to or loss of the premises.

Not every property is suitable for such a lease and each lease will be the outcome of negotiations between the parties. The owner of a large development may insist on the use of a standard lease throughout in which case the prospective tenant is faced with the decision either to accept it or to decline to take a unit on that development.

Most leases are unique and no assumptions should be made concerning their contents. The valuer has no short cut; it is necessary to read the whole of the lease in the context of the statutory background and in relation to the formidable body of case law which has grown up in recent years. It is not proposed to deal in detail with the law but it is pertinent to highlight those areas which affect value, especially where there has been considerable court activity likely to influence market value.

In addition to the repairing covenant and the provisions for rent review, the landlord will seek to exercise some control over the use of the property and to limit the tenant's ability to carry out work or assign or sublet his interest to another.

Specification of the user clause should receive careful consideration to avoid any lowering effect on the rental value which may result from a strict or narrow user provision. In one case, the requirement to use offices only for the work of consulting engineers led to a restriction of the rent imposed on review (*Plinth Property Investments Ltd* v. *Mott Hay & Anderson*). The landlord of business premises where such a user clause has an adverse effect on rent may seek to change the provision on a renewal under the 1954 Act. But without the agreement of the tenant he is unlikely to be successful (*Charles Clements (London) Ltd* v. *Rank City Wall Ltd*).

Where the tenant of business premises wishes to carry out improvements the landlord can offer to do the work himself in exchange for a reasonable additional rent but otherwise will find it difficult to prevent the tenant from himself doing the work. Where the tenant undertakes the work, he will benefit from the provisions of the Landlord and Tenant Act 1954 on renewal, whereby the effect of the improvements is disregarded in setting the new rent. However, the situation on review will depend on the provisions contained in the lease. In one case, an unfortunate tenant found himself paying rent for improvements which he had recently financed as an addition to a rebuilding following a fire (*Ponsford* v. *HMS Aerosol*).

The wish of a tenant to assign his lease is often opposed by a landlord who may employ delaying tactics. Recent legislation (Landlord and Tenant Act 1988) has shifted the balance, enabling the tenant to proceed where the landlord has not responded to his request within a reasonable time. The landlord will often prohibit subletting on the grounds of deterioration in value, possible extra costs of management and the risk of additional health and safety requirements being imposed.

Whilst the repairing covenant under a full repairing lease places considerable burdens on the tenant including the responsibility to renew parts of the structure where necessary, it does not extend to completely renewing a failed building.

It has been described as 'always a question of degree whether that which the tenant is being asked to do can properly be described as repair, or whether on the contrary it would involve giving back to the landlord a wholly different thing from that which he demised. In deciding this question, the proportion which the cost of the disputed work bears to the value or cost of the whole premises, may sometimes be helpful as a guide' (*Ravenseft Properties Ltd* v. *Davstone (Holdings) Ltd*).

In this case the tenant was held liable to pay for a new method of fixing concrete cladding and to provide expansion joints not previously provided since the work proposed was the only way of remedying the defect. But in another case where a house suffered from severe condensation problems it was held that there was no evidence of lack of repair. The cause of the complaint was a design fault to remedy which would amount to an improvement (*Quick* v. *Taff Ely Borough Council*).

Provisions for rent review have become more sophisticated and complicated in recent years alongside the landlord's realization that regular increases in rent are an integral part of the value of the property. Many leases lay down stringent provisions in the machinery for undertaking reviews, neglect of which may seriously disadvantage the tenant. Some of the provisions create unreal situations where the rent is to be assessed in hypothetical circumstances unrelated to the market. When, usually too late, the effect of the provision is realized there is often tension in the landlord–tenant relationship.

One such case, reported in 1985, concerned the interpretation of a rent review clause where it appeared possible that the intention was to fix a rent for the remainder of the unexpired term of 25 years as if no further reviews would take place, although the reality was that the lease provided for rent reviews at 5-yearly intervals. It was obvious that the absence of reviews would result in a higher rental value. In fact the court determined that the rent should be fixed on this assumption. The next review having become due, the tenants have won the right to relitigate the point on the basis of subsequent developments in case law (*Arnold* v. *National Westminster Bank Ltd*).

It is a source of some surprise that tenants continue to agree upward-only rent reviews which may result in the imposition of reviewed rents above the level of the market. So far there have been few cases where this restriction has made any difference to the rent level although in some sections of the office market tenants have found it necessary to offer a lump sum (a reverse premium) to attract an occupier or to persuade an owner to accept a surrender of the lease. From an investor's point of view, rents in excess of the intrinsic rental value add a further risk dimension to property investment.

The landlord will be anxious to be consulted before any alterations are carried out. In many cases, work which the tenant proposes to undertake will be of long-term benefit to both parties although specialist adaptations may reduce the letting value in the market. The landlord's ability to resist improvements proposed by the tenant is much constrained by legislation as noted earlier.

Wherever a dispute arises the result turns upon the facts which are unlikely to be the same in two different cases, leading to uncertainty as to the outcome of any case going to court. Litigation is expensive and it may be that an apparent flaw will be exploited by one party because the other party is unable or unwilling to risk a court case. On the other hand, investors with large land holdings may take what in themselves are quite trivial cases in order to establish, or at least to clarify, a particular principle.

The valuer needs sufficient understanding of the law of landlord and tenant to be aware of the significance of particular provisions, especially where they have been the subject of interpretation by decided cases.

The valuer is not trained in legal interpretation and where he is uncertain as to the precise meaning of a clause, the result of which is likely to have an effect on his view of value, he should seek expert legal opinion and advice.

6.6 SERVICE CHARGES

6.6.1 Introduction

Under the provisions of many leases the landlord maintains closer control of the work for which the tenant is responsible by managing that work which is then charged to the tenant together with a management fee. The procedure is particularly useful and probably essential where the premises are let to a number of tenants.

Services provided to tenants of business premises are administered and charged in accordance with the contractual arrangements agreed between the parties as contained in the lease. Residential tenants were thought to need some protection from the sometimes harsh contractual terms and statute has interceded to regulate the relationship. Charges may be apportioned between the various tenants in relation to rent, rental value, rateable value or floor area.

The differences are sufficient for the two codes to be described separately.

6.6.2 Commercial premises

Service charges on commercial premises are determined by the provisions of the lease. There is no overriding legislation, as is the case with residential premises.

Lease stipulations for the range, provision, apportionment and collection of service charges become very complex in the case of an office block in multiple occupation or a retail shopping development. It is important for the arrangements to be clear and the implementation efficient. The modern lease usually lists a comprehensive range of services controlled by the landlord with further unidentified services to be provided where found to be necessary.

The landlord will normally enter into a number of contracts for items such as cleaning and refuse collection and should ensure that the terms of the contract are complied with. He may decide to deliver certain services directly through staff employed by him, for example, security and landscape maintenance.

The lease will set out the method of cost apportionment. Each tenant's share of the total cost of services may be related to the floor area, a weighted floor area to reflect different requirements, rateable value, retail price index or rent paid. A tie-in to any datum point outside the control of the landlord (for example, rateable value or retail price index) is best avoided as changes may have unexpected and unwanted effects. The collection of service charges usually takes place in two parts. An estimate is produced of the total costs for the year on the basis of which the landlord collects advance quarterly payments, the balance being payable once the actual costs are known and certified.

It is important for estimated costs to be paid in advance, otherwise the landlord will need to have his own funds available to finance expenditure until tenants' payments are received (he may not be able to reclaim any interest charges incurred unless specifically provided for in the lease). There are many unsatisfactory service charge provisions in existence. The result may be that the landlord is unable to collect the whole of his expenditure or that he has to provide at his own expense the working capital necessary to deliver the service.

6.6.3 Residential premises

A report prepared at the instigation of the government showed that lease provisions were often unsatisfactory and sometimes unfair to tenants and that legislation was required to restore a balance between the parties. The result was the Landlord and Tenant Act 1985 as amended by the Landlord and Tenant Act 1987.

The former Act defined a service charge as an amount payable by a tenant of a dwelling as part of or in addition to rent: which is payable, directly or indirectly, for services, repairs, maintenance or insurance or the landlord's costs of management and; the whole or part of which varies or may vary according to the relevant costs.

Improvements will not be included in the definition except where expressly referred to in the service charge provisions of the lease. Some improvement is, of course, incidental in most repairs.

The 1985 Act contains provisions as to the reasonableness of the costs and of the standard of execution of the works. Consultation with the tenant or a recognized tenants' association is required, where the relevant costs exceed £1000 or £50 for each dwelling concerned, whichever is the greater, and the work shall not be started earlier than 1 month after the consultation procedure is commenced.

The tenant or the association may require a written summary of costs, certified by a qualified accountant where the charges relate to payments by tenants of more than four dwellings. There are provisions for subsequent inspection of accounts receipts and other documents relating to the summary.

Service charge payments are required to be held on trust to the exclusion of any other provisions contained in the lease. Service charges payable to a local authority, new town corporation or the Development Board for Rural Wales are not affected by the provisions, unless the lease is for a period of 21 years or more.

6.7 THE BUILDING

The investor will prefer a building which is recently constructed of first class materials, of a design which has aesthetic appeal and the use of which is adaptable to possible future changes in demand. Most important, he will look for a building to be in an appropriate location for its particular use.

Where repairs become necessary, it should be possible to carry them out speedily, without undue expense, and without heavy reliance on specialist proprietary parts which may become difficult to obtain over the life of the building. This appears to disfavour many 'systems' buildings, so popular in the 1960s but which often now require major rehabilitation to enable their continuing use.

The frequency and cost of repairs is important, even where the tenant is responsible for such work. Not only will a building known to be expensive to maintain be more difficult to let or re-let, but any present or prospective occupier will examine the likely level of costs of repair when considering the rent he is prepared to pay, since he is more concerned with total occupation costs than the component parts of such costs. The level of costs will also be a factor in rent review negotiations.

The foregoing specification is one of perfection which few buildings will attain; the point is that any departure from the ideal is a factor against the building. A few minor shortcomings may be acceptable but several more important deficiencies are likely to narrow the potential markets both for tenants and purchasers with a consequent effect on rental and capital values.

The site and its location are important. The valuer will wish to judge the suitability of the site for the building erected on it. He will consider whether it is large enough and caters for any particular requirements associated with its use; whether it has adequate off-site car parking provision and suitable loading and unloading facilities. The relevance of the location to its use will always be important and under some circumstances critical.

The location of shop premises is limited to particular, relatively small, central areas and the difference of a few metres can make a substantial difference in value. Office buildings tend to develop in clusters; one part of the town or one street in the town often gains a reputation as the location for the majority of law firms or accountants which will act as a magnet to others in the same or associated professions. Public transport and or accessible on-site car parking are desirable, whilst staff recruitment may be helped by the proximity of shopping facilities.

Factories need good road access and proximity to a sufficient workforce, although their precise location is rarely of prime importance. Retail warehouses rely on a prominent main road position with good access and ample parking. Retailers and shoppers show a preference for a purpose built retail park housing a number of suppliers rather than the earlier single site.

6.8 THE TENANT'S 'COVENANT'

The financial standing of the tenant is important for trouble-free financial relationships. Where the tenant is a major commercial concern, probably a public limited company with shares quoted on the stock exchange, the rent is likely to be more secure than from a recently formed expanding firm suffering cash flow problems and without ready access to outside funds. The advantages of payment of rent by direct debit on the due date is a feature that can be appreciated fully only by those who have experienced the frustration and time involvement of trying to collect rent from unsatisfactory tenants. The clear implication is that the tenant should be selected with great care and it is part of the valuer's concern to assess the quality of the tenant's covenant to pay rent and perform his other obligations under the lease.

The good tenant will not only pay rent on time but may be more likely to honour repairing and other obligations. The cost of management is thus minimized and the attraction of this type of tenant is such that prospective purchasers may be prepared to accept a slightly lower yield.

New developments are a particular example of the benefit of a known tenant. Where the developer of a shopping centre is able to let space to a major company as an 'anchor' tenant confidence is created which often results in the attraction of other tenants. The large retailing groups are aware of their power and use it to negotiate attractive initial terms in such circumstances.

Any investor is first concerned to know the minimum net income he can expect from a prospective investment. Because of the nature of real estate, that exercise is in itself of a different complexity than it is in the case of government stock or shares, where the dividend is received free of outgoings and in most cases tax has been deducted from the payment. In the case of income from property, the income may be net of outgoings or it may be subject to outgoings for which the owner is liable. There will be regular charges for management where an agent is employed and further legal and other professional charges may be incurred where the rent is reviewed or the property re-let. Value Added Tax may or may not be payable on the rent; where it is, the owner will have a duty to account to H.M. Customs and Excise. These matters will be explained in more detail shortly, but enough has already been said to show that estimation of net income requires knowledge and experience.

6.9 BUSINESS TENANCIES

6.9.1 The impact of statutory protection

Most investments in property involve occupation for business purposes. Tenants enjoy considerable security of tenure; in those cases where possession may be obtained against the wishes of the tenant, statutory compensation is normally available. Either event has implications for the valuation process.

There is a public interest case for imposing limitations on the landlords' ability to obtain possession of business premises where the tenant may have invested considerable time, skill and funds in creating goodwill (which cannot readily be translated or compensated).

The Landlord and Tenant Act 1927 sought to give tenants some protection but was largely unsuccessful and was repealed (except for the part concerned with tenants' improvements) by the Landlord and Tenant Act 1954 (now the principal Act though modified in important respects by the Law of Property Act 1969).

The application of the Act is limited to those tenancies falling within the definition provided by section 23(1).

'Subject to the provisions of this Act, this Part of this Act applies to any tenancy where the property comprised in the tenancy is or includes premises which are occupied by the tenant and are so occupied for the purposes of a business carried on by him or for those and other purposes.'

A business is defined as including

'a trade, profession or employment and includes any activity carried on by a body of persons, whether corporate or unincorporate' (s.23(2)).

There are limited provisions for opting out under the provisions of section 38(4):

'The court may
- (a) on the joint application of the persons who will be the landlord and the tenant in relation to a tenancy to be granted for a term of years certain which will be a tenancy to which this Part of this Act applies, authorise an agreement excluding in relation to that tenancy the provisions of sections 24 to 28 of this Act; and
- (b) on the joint application of the persons who are the landlord and the tenant in relation to a tenancy to which this Part of this Act applies, authorise an agreement for the surrender of the tenancy on such date or in such circumstances as may be specified in the agreement and on such terms (if any) as may be so specified;

if the agreement is contained in or endorsed on the instrument creating the tenancy or such other instrument as the court may specify; and an agreement contained in or endorsed on an instrument in pursuance of an authorization given under this subsection shall be valid notwithstanding anything in the preceding provisions of this section.'

Certain types of lettings are excluded by the provisions of section 43 including agricultural holdings and mining leases. Tenancies of licensed premises have recently been included by the provisions of the Landlord and Tenant (Licensed Premises) Act 1990 although it should be noted that many public houses are managed or subject to a licence agreement which will leave the premises outside the provisions of the 1954 Act. The words used in section 23(1) must be considered carefully. The protection is available to tenants (the Act does not extend to licensees) who occupy the premises for business purposes (where a tenant gives up occupation he no longer has the protection of the Act even though he remains the tenant). If he sublets the premises, the subtenant would be eligible for protection (subject to satisfying the other provisions of the legislation). There is a wide interpretation of the expression 'business'. The parties may agree to opt out but such an arrangement is valid only where both parties make a joint application to the court before the start of the tenancy. According to a Law Commission report, some 15% of applications are refused.

The implications of a 'contracted out' lease are severe in that at the end of a term the tenant will be unable to claim a new lease under the Act; even where the landlord is willing to offer a new lease the tenant will be unable to invoke the powers of the court to adjudicate on those terms where the parties are unable to reach agreement.

The rental value of the premises without the general security of tenure afforded by the Act will tend to be lower than similar premises with the benefit of the Act; that tendency may be reflected where an arbitrator is called upon to fix a rent on review. It should be noted that the parties do not opt out of the whole of the Act but only sections 24–28 relating to:

1. Continuation of tenancies and grant of new tenancies (s.24);
2. Rent while tenancy continues by virtue of section 24 (s.24A);
3. Termination of tenancy by the landlord (s.25);
4. Tenant's request for a new tenancy (s.26);
5. Termination by tenant of tenancy for a fixed term (s.27);
6. Renewal of tenancies by agreement (s.28).

Even though the tenancy is for a fixed predetermined term of years (a term certain), the tenancy will continue until determined at its expiration or later by one or other of the parties to the lease.

Where the tenant wishes to leave, he may give a notice in accordance with the terms of the lease (but he must give not less than 3 months notice to expire either at the end of the term or on any subsequent quarter day as provided in s.27).

Should the tenant wish to stay but be anxious for the grant of a new term of years, he may make application as provided in section 26, serving notice in the form prescribed by the current Landlord and Tenant Act 1954 Part II (Notices) Regulations 1983 (SI 1983/133) and the (Notices Amendment) Regulations 1989 (SI 1989/1548).

Where the landlord decides to serve a notice, either because he wishes to obtain possession or simply because he wishes to renegotiate the terms, the notice must be in the form prescribed in the regulations as provided by section 25. Where the landlord is willing to grant a new lease he must say so; if he is not willing to do so, he must specify the reason or reasons for not doing so which must fall within the provisions of section 30(1) (a)–(g) which may be summarized as follows:

(a) Breach by the tenant of the repairing covenant;
(b) Persistent delay in paying rent due;
(c) Other substantial breaches by the tenant of his obligations;
(d) Alternative accommodation provided or secured for the tenant by the landlord (it must be suitable, reasonable and available at a time to

suit the tenant's requirements including the requirement to preserve goodwill);

(e) Uneconomic letting. Where the current tenancy was created by a subletting of part only of the property and which where the aggregate of rents on separate lettings produces a rent substantially less than the rent reasonably obtainable on a letting of that property as a whole;

(f) Demolition of reconstruction of the premises or a substantial part of them. But by section 31A (introduced by the Law of Property Act 1969) the tenant agrees in the terms of the new tenancy to the landlord having access to carry out the work and that it could be carried out without obtaining possession or that the tenant is prepared to accept a tenancy of an economically separable part of the holding;

(g) Occupation by the landlord when the current tenancy comes to an end, either wholly or partly for a business carried on by him there or as his residence. This ground applies only where the landlord's interest was purchased or created five years or more before the termination of the current tenancy.

Various decisions of the courts have underlined that substantial evidence will be required of conduct alleged in grounds (a)–(c) and which, if proved, will deprive the tenant of occupation and any compensation to which he would otherwise have been entitled.

Where suitable alternative accommodation is offered (ground (d)) the tenant is not entitled to compensation, underlining the fact that the compensation is intended to compensate for loss of goodwill (a word used in the Act but not defined). It follows that one of the tests of suitability in this case is whether the tenant can retain his existing goodwill in the alternative premises offered.

The remaining grounds (e)–(g) will, if proved, entitle the tenant to compensation based on the rateable value and the period for which all or part of the premises have been occupied for business purposes and on any change of occupier each succeeded to his predecessor's business. The rateable value multiplier is 8 for periods of less than 14 years and 16 for periods of 14 years or more based on the 1990 Valuation Lists. The tenant has the option to elect for a multiplier of 1 (less than 14 years) or 2 (14 years or more) based on the 'old' (1973 list) valuation. The details are complex and reference should be made to the Landlord and Tenant Act 1954 (Appropriate Multiplier) Order 1990 (SI 1990/363).

Either multiplier represents a substantial sum and the landlord will wish to consider the implications, for example, in ground (f) where he proposes to carry out reconstruction work where the tenant may seek to stay in the premises whilst the work is being done in accordance with provisions in section 31A(1). The landlord would be absolved from paying compensation and may wish to consider whether it is possible for the work to be done in this way.

Finally, where the landlord requires possession for his own occupation (whether to use as business premises or for these and other purposes) he must be able to show ownership for a minimum period of 5 years (s.30(2) and (3)).

Notice under section 25 of the Act requires a response within a specified time which cannot be extended so the tenant should act quickly in order to protect his rights.

Tenants benefit greatly from any delay in settling the new rent because the rent is not backdated on settlement to the end of the previous tenancy (though any increase in market rent during that time may be reflected in the rent fixed). Section 24A was added to the 1954 Act by the 1969 Act to enable the landlord, by notice, to claim an interim rent the effect of which is to give the opportunity for a revised rent from the beginning of the notice which greatly reduces any benefit of delay to the tenant. It is important to serve the notice early in the proceedings since any higher rent cannot be backdated and is payable only from the date of the notice. It does not eliminate all benefits, since the interim rent is based on a rent from year to year having regard to the existing tenancy which tends to produce a rent below market value by perhaps 10–20% according to decisions reached by the courts. However, the interim rent so determined is payable whether or not the tenant eventually proceeds with a new tenancy so the tenant takes some risk in staying on without knowing the level of rent to be fixed.

The parties negotiate for a new tenancy where they both wish to continue or where the landlord's reasons for terminating are not upheld by the court. Either party may have recourse to the court where agreement cannot be reached but it is important to note that where the parties have reached partial agreement they may request the court to complete the agreement by ruling on the outstanding items only, incorporating the previously agreed items in the order of the court for a new tenancy.

Three sections of the Act relate to the terms and conditions of the new tenancy. Section 33 provides for a maximum duration of 14 years and the court is empowered to include provisions for earlier review. However, whilst this limitation binds the court, a partial agreement by the parties for a longer term could be incorporated in the court order should they so wish.

Section 34 sets out the 'disregards' in relation to the market rent to be fixed whilst section 35 relates to any other terms in the lease.

Apart from the compensation referred to earlier, there is a provision for compensation for misrepresentation (s.55). Some writers suggest that any compensation would be nominal but it seems more likely that the absence of a ceiling would enable the tenant to itemize and claim for loss incurred either in transferring or closing his business which could result in a very substantial claim.

Finally, the tenant is entitled to compensation for improvements carried out by him following the procedure laid down by the 1927 Act. He must have obtained the landlord's consent to the work or a decision of the court before commencing the work; the landlord is entitled to elect to carry out the work if he so wishes and to charge a reasonable rent therefore but he cannot insist on providing the improvement where the tenant decides not to proceed.

Where the tenant carries out the improvements, compensation when his tenancy terminates may be substantial as calculations are based on values and costs current at the time of the claim. The compensation is the lesser of the net addition to the value of the property as a whole or the reasonable cost of carrying out the improvement at the termination of the tenancy less any costs of repair.

It will be appreciated that the tenant is unlikely to receive any compensation where the site is to be cleared and redeveloped, a consideration that may guide the landlord when deciding at an earlier stage whether to elect to provide the improvements proposed by the tenant. The county court has jurisdiction in these matters where the rateable value of the property does not exceed £5000, the High Court dealing with properties having a higher rateable value. The parties may agree in writing that jurisdiction can be transferred from the county court to the High Court or from the High Court to a county court specified in the agreement (s.63(3)).

6.10 RESIDENTIAL TENANCIES

Residential property has a long history of statutory interference with the contractual relationship between the parties. Rent and mortgage interest restrictions were initially imposed during the First World War and most tenancies have been subject to some form and level of control since that time. Other parts of the legislation were directed towards giving security to tenants and rights of succession to certain members of the family.

Many landlords reacted by selling when the opportunity arose rather than re-letting. This was sometimes because the opportunity to realize a capital sum was attractive but was also influenced by the cost of maintaining property, particularly the older types, leaving only a nominal net income.

The serious shortage of accommodation was alleviated to some extent by a large public provision of council housing between the wars and for some time thereafter. The accommodation was of a better standard than most of the tenanted accommodation in the private sector and in particular provided bathrooms, indoor sanitation and a garden. The waiting lists were long however, and not everyone qualified for acceptance on to the lists. Few, if any, houses to let were built by private investors and with the contraction of council house building and the recent 'right to buy' provisions, the total stock has declined.

Attempts to make more accommodation available in the private sector by rewarding the landlord with a 'fair rent' defined to exclude scarcity were largely unsuccessful.

With the enactment of the Housing Act 1988 a new era commenced. It provided that many tenancies granted after 14 January 1989 will have the more oppressive restrictions of earlier legislation lifted; the landlord is able to charge and maintain a market rent and resumption of possession is facilitated, although there is security of tenure for the tenant who observes the terms of the letting (possession is available to the landlord only on certain specific mandatory or discretionary grounds).

As has been seen with business tenancies, obtaining possession is not of major importance to the investor provided he is able to maintain a rent level which compensates for inflation and gives a proper return on capital. The Act introduces a new form of assured tenancy (the grant of which is not limited to approved landlords as was the case under earlier legislation). It also introduces the assured shorthold tenancy under which the landlord may obtain possession on giving notice to expire on or after the end of a fixed term.

The new legislation should encourage some investors to offer houses to let, having acquired them specifically for that purpose, possibly taking advantage of the Business Expansion Scheme to make the investment more attractive. There are two principal reasons why investors may not be too enthusiastic; rent controls could be introduced by another administration and the unit cost of management is relatively high.

6.11 THE PLANNING FRAMEWORK

Development control is assured by the requirement that planning permission is required for the

'carrying out of building, engineering, mining or other operations in, on, over or under land or the making of any material change in the use of the buildings or other land.' (s.22 Town and Country Planning Act 1971.)

The General Development Order 1977 (as amended) and the Use Classes Order 1987 allow certain work or changes of use to be effected without permission.

Development takes place where the change of use is material although the following changes of use are excluded from the definition of development:

1. Use of the buildings or land within the curtilage of a dwelling house for any purposes incidental to occupation;
2. Use of any land for forestry or agricultural together with any buildings on the land;
3. Use of land or buildings for any purpose specified in an order made by the Secretary of State.

Change of use within a use class does not constitute development whilst change from one use class to another does; recent changes resulting from the Use Classes Order 1987 have widened the scope of some of the classes which are:

Class A1 Shops
Class A2 Financial and professional services
Class A3 Food and drink
Class B1 Business
Class B2 General industrial
Class B3 Special industrial Group A
Class B4 Special industrial Group B
Class B5 Special industrial Group C
Class B6 Special industrial Group D
Class B7 Special industrial Group E
Class B8 Storage and distribution
Class C1 Hotels and hostels
Class C2 Residential institutions
Class C3 Dwelling houses
Class D1 Non-residential institutions
Class D2 Assembly and leisure

Structure plans and local plans together comprise the development plan which gives guidance as to the policy of the local planning authority. There are detailed provisions for making application for planning permission including certification of ownership or notification of owners and information to neighbours where the application involves development of a 'bad neighbour' nature or where the property is situated in a conservation area. Where an application for planning permission is refused, granted subject to conditions or where no decision is reached within a certain time, where an enforcement notice is served or the authority fails to issue an established use certificate the applicant may appeal to the Secretary of State for the Environment. The appeal may be heard at a public local inquiry but the parties may agree to proceed by way of written representations.

Special provisions apply to the following:

Listed buildings
Conservation areas
Tree preservation orders
Caravan sites
Advertisements
Scheduled monuments
Areas of archaeological importance
Sites of Special Scientific Interest
National Parks
Enterprise Zones
Simplified Planning Zones

Detailed planning enquiries should always be made prior to the formulation of a valuation since the information obtained may have a fundamental effect on value.

There are detailed regulations to enable the enforcement of planning control and any unauthorized development, breach of a condition of planning permission or use for an unauthorized purpose may result in service of an enforcement notice.

7 | The valuation process

It is only following the gathering of information, the exhaustive market investigations referred to earlier and a consideration of the nature of the legal interest that the valuer is in a position to proceed with the numerical valuation.

7.1 CONVENTIONAL VALUATION APPROACH

Probably the majority of practising valuers still prefer to value investment properties using implicit all risks yield, notwithstanding the inability to make direct comparisons of returns available in the wider investment market.

The conventional or traditional approach ignores anticipated future increases in rental value. The current rental value is treated as the income likely to be received indefinitely once the current rent can be adjusted even though one of the main attractions of investment in property to most investors is the likelihood of regular increases in rental value. This distortion of expected performance is then counterbalanced by using an all risks yield which is below current yields available on other types of investment, in anticipation of regular and substantial growth which is not quantified.

The yield is derived from market evidence with its inherent reflection of growth potential but is then adjusted intuitively to reflect differences in the property to be valued, such as quality of tenant covenant, lease provisions including rent review pattern and the suitability of the location. Initial, current and reversionary yields are examples of all risks yields which may be calculated for any investment where the current rent is below market value. None will necessarily give a true reflection of the market view of the all risks yield which is derived from analysis of market transactions, preferably of rack rented investment. The relative merits of the implicit and the explicit approaches to valuation will be examined later.

7.2 RACK RENTED INVESTMENTS

The rack rented investment property presents the most straightforward case. Where the rent is at full market value and the property is of freehold tenure let on modern lease terms and in particular with regular rent reviews at not more than 5-yearly intervals, the procedure involves capitalizing the rent in perpetuity at an appropriate yield derived from market observation and analysis. In other words, the right to receive a flow of income at the current level in perpetuity is discounted at a market derived all risks yield to find the present net value.

Example 7.1

Value a modern freehold office building occupied by one of the major banks where the property was let 2 months ago for a term of 25 years with 5-yearly reviews on a full repairing and insuring (FR and I) lease at a rent of £79 000 p.a. Recent transactions in the immediate locality suggest that investors are willing to accept an all risks yield of 6.5%.

The purchase price at which the investment will show a yield of 6.5% is found from R/i

where R = net rent payable
     i = the yield (expressed as a decimal fraction)
Thus R/i = £79 000/0.065 = £1 215 385

The valuation is more often set out in the following form

	£
Net rent	79 000
YP perpetuity @ 6.5% (1/0.065)	15.385
	1 215 385

Commentary
This is the simplest case. The rent is a recent one and is likely to represent the current rental value unless an initial premium has been paid. That rent will be assumed to be receivable in perpetuity. The tenant is of undoubted status and management should be straightforward. Regular 5-yearly reviews give an opportunity to maintain the rent at a proper level.

The question gives some information on market yields achieved in other recent sales, indicating the return expected by investors.

7.3 INITIAL AND ANNUAL COSTS

The investor who purchased on the basis of the valuation shown above may feel in the event that his investment has underperformed since it has ignored certain expenses, both initial and recurring.

The initial transaction will involve legal costs, stamp duty and valuation fees (together with VAT where applicable). The actual costs are not based on a fixed scale but will be influenced by the time taken and the complexity of the case. A reliable average figure is about 2.75% of the purchase price to include stamp duty (currently chargeable at 1% on the total consideration where the purchase price exceeds £30 000).

Once acquired, the investor will expect to employ a surveyor to manage the property — collect rents, ensure the performance of covenants to maintain and repair, check that the insurance premium is being paid and that the building is covered for an adequate sum for all necessary risks, negotiate rent on review (with the possible expense of taking the question to arbitration), monitor financial performance and advise on investment strategy. Taking one year with another, the average annual management cost in this example may be in the region of 1.5% of rent collected.

The valuation can now be refined.

Example 7.2

	£
Rent reserved (net)	79 000
deduct management expenses @ 1.5%	1185
	77 815
YP perpetuity @ 6.5%	15.385
	1 197 184
deduct transaction costs of 2.75%	32 041
	1 165 143

Commentary
The result is sufficiently different to suggest that such costs should be reflected. But this requires that the comparable evidence has been dissected and analysed in the same way. Analysing and valuing on a gross basis (i.e., ignoring annual management expenses and costs of acquisition) would avoid the tedium of the additional calculations. The problem is that the expenditure is not necessarily constant between one transaction and another.

The transaction costs are calculated on the net valuation, the figure of £32 041 representing 2.75% £1 165 143.

It is usual to round down, in this case to £1 165 000 or perhaps more likely to £1 160 000. To report a more precise figure would give a wrong impression of the degree of accuracy achievable on such a valuation exercise.

The valuer would also include some advice and possibly suggest a range within which it would be worthwhile to consider purchase. For example, he may complete his report with a recommendation:

'In our opinion the current market value of the above described unencumbered freehold interest subject to confirmation of tenure, lease terms and other assumptions stated is £1 160 000 (one million one hundred and sixty thousand pounds).

In our view the asking price of £1 400 000 is in excess of the current market value of the property: we believe that purchase at a price of up to £1 200 000 would represent a satisfactory investment acquisition.'

It will be noted that this upper figure (not a valuation but negotiation advice) is within a range of approximately 5% and below the original valuation prepared without allowing for initial costs and annual expenses. In the valuation examples that follow adjustments for costs of acquisition and management will be omitted.

Each example is intended to demonstrate a particular point which will be clearer without the intrusion of information not directly relevant to valuation principles.

There seems to be a general reluctance to use yields other than in multiples of 0.5%, probably because published tables offer only a limited number of steps in the interest rates available. With the advent of suitable calculators there is no such limitation on available rates; any desired rate my be used in a calculation. Yields on stocks and shares are calculated to the nearest one-eighth per cent and it seems reasonable to endeavour to view property investments in the same way.

7.4 THE TWO-INCOME MODEL

So far, the examples have been concerned only with those cases where the tenant is paying the full market value.

The majority of investment valuations are less convenient than this, containing two or more income levels. The first, payable for the remainder of the current term (that is, until the next rent review or lease renewal) is a rent which was fixed 1, 2 or more years ago and is therefore likely to be below the current rental value. Should that be the case, the current rental value should be estimated, using whatever assistance can be gained from information regarding recent lettings of similar properties. There is no

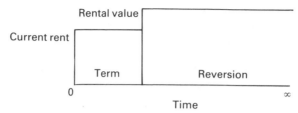

Figure 7.1 Term and reversion (vertical division) model.

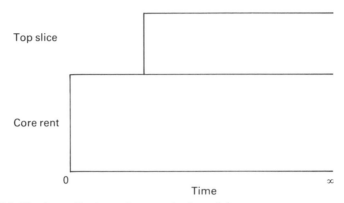

Figure 7.2 Hardcore (horizontal separation) model.

unanimity about the treatment of the varying incomes for valuation purposes within the conventional approach.

Some valuers prefer to treat the income as comprising two separate
vertically divided blocks (distinguishing between the current rent for the
remainder of the term and the anticipated rent payable thereafter) whilst
others treat the current rent as continuing into the future with a separate
horizontal layer representing any anticipated increase. The different treatments are shown diagrammatically in Figures 7.1 and 7.2.

Vertical and horizontal models may be operated using the same yield
throughout (equivalent yield) or the yields may be varied to reflect
perceived differences in risk between the two parts, giving four treatments in total which will now be described.

7.4.1 Term and reversion (vertical division) approach

The earliest model is the two-yield vertically divided term and reversion
approach. The main (reversionary) income will normally account for by
far the larger part of the capital value. That tranche of income is capital-

ized using an all risks yield derived from an analysis of sales of comparable but rack rented property where such information is available. The lower term rent is then capitalized at a slightly reduced yield (typically 0.5–1%) to reflect the greater certainty and security in that part of the income.

The application of each model to be described may be demonstrated by use of the same basic information, relating to a freehold shop property currently let on a modern institutional type lease at a net rent of £41 000 p.a. has 2 years to run to the next rent review. The current rental value is £50 000 p.a.; market evidence suggests a yield of 5.5%.

Example 7.3

Term and reversion — variable yield

Term

		£
Current rent	41 000	
YP 2 years @ 5%	1.859	76 235

Reversion

Current rental value	50 000	
YP perpetuity def. 2 years @ 5.5%	16.336	816 800

Capital value		893 035

Commentary

The current rent will continue for 2 years when a new rent will be negotiated. Note that the figure used is not a projection of what rent is likely to be payable in 2 years time but the estimated current rental value. The expectation that the rental value will continue to increase is reflected in the low all risk yield used.

The capitalization of the term income takes place at 5% as opposed to the yield for a similar rack rented property of 5.5%. The lower yield is intended to reflect the greater security of the current rent which is below market value, as well as the certainty afforded by the existence of a contract between the parties. In contrast, with regard to the future rent the parties have done no more than agree to renegotiate it at particular intervals; there is not the same quality of certainty about that income. Against that, although the rent will not be renegotiated for 2 years when rental values are expected to have increased, the current level of rental value is used.

Example 7.4

Term and reversion — equivalent yield

Term

		£
Current rent reserved	41 000	
YP 2 years @ 5.5%	1.846	75 686

Reversion
Current rental value	50 000	
YP perpetuity def. 2 years @ 5.5%	16.336	816 800

| Capital value | | 892 486 |

Commentary

The only difference between Examples 7.3 and 7.4 is in the value of the term income, which is £549 less in this case. Not only is it very small but it is insignificant in terms of the accuracy expected of market valuations. Notwithstanding that, both treatments have their adherents and it is worthwhile to look at the reasoning employed.

In the two-yield model, it is argued that where the investor is prepared to accept a certain yield as evidenced by market transactions, any part of the income below current rental value represents a lesser risk and would therefore be welcomed as part of the total investment and discounted at a marginally lower yield to reflect the security.

The weakness of the argument seems to be that there is no way in which comparable transactions can be analysed to expose such reasoning on the part of the investor. It may seem a natural reaction to reduce a yield where the risk is less but there is no evidence that the market adopts such an approach. The risk rate is attributable to the property and not to a part of it. This approach also destroys the claim that it mirrors the market as there is no evidence that the market operates in this way. Where the market has indicated a preference for a particular yield, the valuer should reflect carefully before ignoring that preference by making an adjustment to one part of the income flow, even though in general the difference in capital value is insignificant. The market yield may be adjusted where the property to be valued is different in significant respects from those providing the market evidence.

The equivalent yield model may be presented in a slightly different way.

Example 7.5

			£
Current rental value	50 000		
YP perpetuity @ 5.5%	18.182		
			909 100
less marginal income (£50 000–41 000)	9000		
YP 2 years @ 5.5%	1.846		16 614
Capital value			892 486

Commentary
The result is the same as that found in Example 7.4 but the procedure has some advantages. It draws attention to the potential of the investment by showing the maximum market value when let at the full market rental value in perpetuity. It deducts the shortfall for the next 2 years to find the lower capital value due to the current rental level. It should be slightly easier to visualize, leading to a clearer approach to valuation.

7.4.2 Layer (horizontal separation) approach

Equivalent yield
The vertical model previously described reflects the way in which the income is anticipated and takes account of the change to a higher income when it happens. The variation now described adopts an arbitrary position whereby the current income is assumed to continue into perpetuity, the increase being treated as a separate stream in perpetuity but commencing at some time in the future. As a result there is an artificial horizontal separation of core and marginal or incremental incomes.

Layer — variable yield
The true hardcore model uses two yields and will be described shortly. The model more frequently used is the equivalent yield model which may be performed in either of the following ways.

Example 7.6

		£
Current rent	41 000	
YP perpetuity @ 5.5%	18.182	745 462
increase on reversion to rental value	9000	
YP perpetuity def. 2 years @ 5.5%	16.336	147 024
Capital value		892 486

Example 7.7

		£
Rental value	50 000	
YP perpetuity @ 5.5%	18.182	909 100
less shortfall for 2 years	9000	
YP 2 years @ 5.5%	1.846	16 614
Capital value		892 486

Commentary

A comparison of the results of Examples 7.6 and 7.7 with the results of Examples 7.4 and 7.5 will show that where an equivalent yield approach is used the result will be the same whether carried out by vertical division or horizontal separation.

It is claimed that the horizontal separation or layer method is more often used, even though this requires an arbitrary and unreal division of the income in the reversionary period, whereas the amount payable will be an undivided sum.

Individual preferences of valuation approach are heavily ingrained. Even in large organizations there is considerable scope for the individual valuer to follow his own inclinations in valuation methods unless the process is fully computerized.

The arguments would appear to be much in favour of an equivalent yield usage to maintain a direct relationship with market information and a vertically divided model, to ensure retention of a link with the expectations of rental change.

The hardcore model was developed in the 1950s and described in detail by White. It is clear that it was intended to deal with a particular tax situation. The model has gained a certain status despite warnings against its use, particularly in its shortened form, by Trott and others. The true model separates the income horizontally and capitalizes the separate income flows at different yields, the yield for the layer stream being calculated from the market yield.

It is claimed that the approach should be used because future increases are uncertain both in terms of market movements and of possible legal restrictions such as were introduced by the Counter Inflation Act 1973 under which a rent freeze was imposed.

But even where a property is let at full rental value a substantial proportion of any capital value is in anticipation of regular rent increases. A hardcore valuation would not change the result in such a case, since the whole of the income is treated as hardcore. A further danger of the approach is that market yields are of no assistance in determining a yield for the marginal element yet valuers will make a judgement of such a yield, typically adding 2% to the yield used for the hardcore component. The only safe way of making use of the model is to carry out the calculations to find the yield for the marginal income.

Example 7.8

		£
Current rent reserved	41 000	
YP perpetuity @ 5.5%	18.182	745 462
increase in income	9000	
YP perpetuity def. 2 years @ 7.5%	10.717	96 453
		841 915

Commentary
The approach differs from the equivalent yield model in that the marginal income is valued at a higher yield, in this case 2% higher, intended to reflect the relative insecurity of this part of the income. There are two problems with this approach. The higher yield is based on the intuitive response of the valuer which is an unsatisfactory way of estimating risk or reflecting the market. As a result the valuer appears to be suggesting that the absence of any certainty regarding future increases incurs a 'penalty' of some £50 000, despite the fact that the current rental value is used in estimating the value of the top layer, which should be a fairly safe estimate of the rent likely to be achieved.

The example was calculated using the market yield for the core income. If the example is now reworked using a yield slightly lower than the market yield for the core income, as is advocated by some supporters of the hardcore approach, the result becomes

Example 7.9

		£
Current rent	41 000	
YP perpetuity @ 5%	20	820 000
increase in income	9000	
YP perpetuity def. 2 years @ 7.5%	10.717	96 453
		916 453

Commentary
The slight adjustment to the yield used to capitalize the core income results in a valuation in excess of the market value of a similar rack rented property (£909 091).

The volatility of the approach renders it unsafe unless the yield applicable to the marginal income is calculated rather than estimated. The method of calculation involves finding the capital value of the marginal income from which the yield may be determined as shown in the following example.

Example 7.10

		£
Rack rental value	50 000	
YP perpetuity @ 5.5%	18.18	909 091
Current rent	41 000	
YP perpetuity @ 5%	20	820 000
Capital value of marginal rent		89 091

Yield on marginal income of £9000 is

$$\frac{9000}{89\,091} \times 100 = \frac{10.102}{\text{say } 10\%}$$

The valuation may now be performed

		£
Core income	41 000	
YP perpetuity at 5%	20	820 000
Marginal income	9000	
YP perpetuity def. 2 years @ 10%	8.264	74 380
		894 380

Commentary
The result is very similar to the answer to Example 7.4 without the benefit of using market yields. An analysis on this basis would not be helpful in determining market response since the yield for the marginal income reflects merely the coincidence of the period of deferment and the relationship between the current rent and the full rental value.

7.5 LONG TERMS AT FIXED RENTS

There is one type of reversionary valuation where the conventional approach requires that a difference between the yield for term and reversionary yields should be maintained.

Investors accept in general that a property investment is satisfactory if the lease contains provisions for regular rent reviews: 5 years is normally regarded as a satisfactory interval. This raises the question as to the treatment of a valuation where the term rental will continue unchanged for a considerably greater number of years. Such an income may be termed 'inflation prone' and reflects more the characteristics of the fixed interest investment markets than those usually associated with property and income growth. An investor in property would not normally contemplate such an investment. He does so only because property is a long-term investment and the opportunity exists to renegotiate the terms at the end of the existing lease to ensure that inflation is reflected in the future.

The valuation should recognize the fixed non-growth term income by discounting at a market or non-property yield. The effect of this treatment may be seen if the earlier Example 7.3 is adapted by taking the period to the next review as 12 instead of 2 years.

Example 7.11

Term

		£
Current rent	41 000	
YP 12 years at 5.5%	8.619	353 359

Reversion

Current rental value	50 000	
YP perpetuity def. 12 years @ 5.5%	9.563	478 165
Capital value		831 524

Commentary
Although the income may be expected to have increased dramatically in 12 years time (doubled at 6% increase p.a. and trebled at 9.5%) the reversionary income is still restricted for valuation purposes to the current rental value. The main problem is that the fixed income for the next 12 years is being treated as if it will enjoy growth, one of the main reasons for holding property investments.

In this context, the difference of approximately £60 000 between the valuation of a rent which is expected to increase in 2 years time and at regular intervals thereafter and one where the rent is fixed for the next 12 years does not adequately reflect the reality. Further, although the future income may be valued to reflect present initial market yields (anticipating the prospect of a modern lease with regular rent reviews being negotiated at the end of the current lease), the deferment should, it is argued, be at the opportunity cost of money (assuming 12%) to reflect that the investment will not show any growth for the first 12 years.

If the valuation is adjusted to follow this reasoning it becomes:

Example 7.12

Term

		£
Current rent	41 000	
YP 12 years at 12%	6.194	253 970

Reversion

Current rental value	50 000	
YP perpetuity @ 5.5%	18.182	
	909 100	
PV £1 in 12 years @ 12%	0.25668	233 345
		487 315

Commentary

There is a substantial difference in the final valuations in Examples 7.11 and 7.12. It will be noted that the value of the term income in Example 7.12 is approximately £100 000 below the term income in Example 7.11. The value of the reversionary income is affected also, since the cost of that tranche has been provided by the investor without any prospect of growth until 12 years hence and is therefore deferred at the market yield (whilst the opportunity for negotiating a modern lease with regular review intervals commencing in 12 years time is reflected in the capitalization of the reversionary income at the all risks yield).

This may be a case where a discounted cash flow approach using money market yields would be more appropriate. (The hybrid model described in Chapter 8 is also of relevance.)

7.6 APPLICATION TO TERM INCOME

Some observers suggest that money market yields should always be applied to the term income to avoid the anomaly of using low yields to reflect growth when there is no growth in a term income.

The argument overlooks the fact that the market has accepted 5-yearly reviews and that most valuations will therefore include a mild reversionary element. Pursuing this argument, the investor is likely to accept returns at all risks yield levels unless the reversion is deferred for longer than 5 years. An example will demonstrate the problem.

Example 7.13

		£
Current rent	41 000	
YP 2 years @ 12%	1.6901	69 294
reversion to current rental value 50 000		
YP perpetuity def. 2 years @ 5.5%	16.3355	816 775
		886 069

Commentary

The result shows a reduction of £23 000 on the capital value of the same property let at full rental value. Looked at in another way, the penalty exacted by the purchaser for foregoing a full income for 2 years is some £23 000 (for an annual shortfall of £9000).

7.7 LEASEHOLD INTERESTS

So far, the examples have considered the valuation treatment of freehold investments only.

The valuer is often called upon to express an opinion as to the value of a leasehold interest. Where the rent of a leasehold interest is less than the rental value, the difference between the two is referred to as a 'profit rent' in the hands of the leaseholder. The profit rent may be enjoyed for the whole of the term or only intermittently; it may retain a constant relationship where such provision is made by the lease or the relationship may vary primarily due to the lack of coincidence between head and subleases.

Such complicated relationships increase the difficulty of analysis and valuation; possibly a greater difficulty is posed due to the wasting nature of all leaseholds, especially where the interest to be valued is for a term of 25 years or less.

Given the lack of comparable leasehold transactions, the traditional method of valuation has been to adapt a yield from a similar, but freehold, investment. It is argued that the leasehold nature of the investment makes it less attractive than an investment in a freehold property and that this difference should be acknowledged by increasing the yield by one or two points and providing for a notional sinking fund to enable the leasehold investment to be replaced at the end of its life, thus perpetuating the income and justifying the comparison with a freehold interest. The amount paid into the sinking fund is taken from taxed income and provision must be made for a sufficient sum to be available after tax. The yield on the sinking fund is taken at a low, safe rate since the investor is judged to be unwilling to take any risks of non-payment since this would put the future capital payments in jeopardy. A guaranteed level of interest on annual payments over a long term would tend to attract relatively low rates. It is usual to take account of tax payable on interest by using net of tax interest rates. Net rates of 3 or 4% are in most use, since these are the rates available in published valuation tables. Investments with a short life would require the major part of the income to provide a sinking fund.

Example 7.14

An office building is held on lease for an unexpired term of 15 years without review at a rent of £10000 on FR and I terms. The current rental value is £17500 p.a. An appropriate freehold yield on a rack rented property is 9%. Interest rates on safe long-term investments are in the region of 5% gross. The standard rate of tax is 40%. Value the leasehold interest.

	£
Rental value	17 500
Rent reserved	10 000
Profit rent	7500
YP 15 years 11%, 3%, tax 40%	5.010
	37 573

The allocation of the profit rent may be analysed

	£
11% return on purchase price of £37 573	4133
Profit rent	7500
Return on investment	4133
Gross amount for sinking fund	3367
less tax @ 40%	1347
Net amount for sinking fund	2020

and the ability of the sinking fund to replace the original cost may be checked.

	£
Annual sinking fund	2020
Amount of £1 p.a. 15 years @ 3%	18.589
Capital replaced	37 573

Commentary

The problem of course is that the original purchase price is the amount replaced: assuming that yields have not changed, the investor will achieve another investment yielding roughly the same initial level of income. In other words, the purchasing power of the investment has not been maintained and the attempt to adapt a freehold property yield is shown to be unrealistic.

The calculations show both the interest on capital and the replacement of capital. A sinking fund rate of 3% is used which reflects the availability of a safe gross rate of 5% and a tax rate of 40%. A similar income but part of a freehold investment and not therefore needing to be replaced as a wasting asset would have a much higher value.

Example 7.15

	£
Profit rent	7500
YP 15 years @ 9%	8.0607
	60 455

Commentary

The yield is taken at 9% to reflect the accepted differential between freehold and leasehold investments. Even taken at the leasehold rate of 11%, the capital value would be £53 932.

The nature of the discounting process ensures that capital is replaced by reinvestment of the surplus income even though no specific sinking fund is provided. This time the reinvestment rate is assumed to be the same as the remunerative rate, namely 9%:

		£
	Profit rent	7500
	9% return on investment	5441
	Balance for sinking fund	2059
check	Annual sinking fund	2059
	Amount of £1 p.a. 15 years @ 9%	29.361
	Capital replaced	60 455

However, this neglects to take account of the tax payable on that part of the rent diverted to the sinking fund. At the 40% rate of tax used previously, the sinking fund would need to produce a return of 15% to replace the original sum. But if an investment showing such a return is available and suitable as a sinking fund, why should an investor purchase the leasehold interest in the first place?

It has long been known that very few investors take out any form of sinking fund and it seems unlikely that investors do any more than demand a sufficient return from the particular investment, knowing that the income is for a limited period and that there is no residual capital value. Looked at in this way, the interest valued in Example 7.14 shows a return of almost 20% p.a. for 15 years. In general the investor would find the decision between a limited term investment and a perpetual investment much simpler presented in this way; the price of a much higher annual return is the eventual loss of capital.

An alternative to the dual rate leasehold approach is to discount the anticipated flows of income at an appropriate equated yield, reflecting any rental growth expected where reviews exist to take advantage of increased rental values. The rent payable to the freeholder is then treated similarly and deducted from the earlier total.

Example 7.16

Modern shop premises are let on a 21 year lease by the freeholder, without review and with an option on the part of the tenant to renew for a further 21 year term on the same conditions except as to rent. There are four years to run to the end of the original lease which is on FR and I terms subject to a rent of £5000 p.a. The head lessee has recently exercised his option, following which he re-let the premises for the period to the end of the new term acquired at a rent of £25 000 p.a. subject to 5-yearly reviews. The rent payable to the freeholder has been agreed at £32 500 p.a. Initial yields for similar property let on modern terms are in the region of 5%. The equated yield may be taken as 12% from which the implied rate of rental growth is calculated at 7.635%. It is assumed that the leasehold would require a higher yield, say 15%, to compensate for

Years	Amount £1 @ 7.635%	Rent (£)	YP @ 15%	PV @ 15%	PV of income (£)	
Rent received						
1–5	—	25 000	3.3522	1.0	83 805	
6–10	1.4457	36 142	3.3522	0.49718	60 236	
11–15	1.4457	52 213	3.3522	0.24718	43 264	
16–20	1.4457	75 430	3.3522	0.12289	31 075	
21–25	1.4457	108 971	3.3522	0.06110	22 319	240 699
Rent paid						
1–4	—	5 000	2.8550	1	14 275	
5–25	—	32 500	6.3125	0.57175	117 229	131 574
Capital value of head leaseholder's interest						109 125

the additional risks associated with the leasehold interest. The head leaseholder's interest is shown below.

Commentary

The freeholder probably argued that the lack of reviews should be reflected in a higher rent for the new lease recently granted as a result of the exercise of the option. The likelihood of success will depend on the way the lease is worded; an increase of 1.5% for each year in excess of the normal review interval (21 − 5 = 16) would suggest a new rent of £31 000 p.a., although it will be noted that the parties have reached agreement on a fixed rent of £32 500.

The arrangement is a good one for the head leaseholder who receives regular increases in rent whilst not paying any extra amounts to the freeholder. Use of an all risks yield in a tax adjusted dual rate approach would not reflect the advantages of this particular interest to the head leaseholder. He holds an advantageous lease where the rent changes only once (in 4 years time). There follows 1 year in which he shows a net loss; thereafter he has a rising profit rent with no further review of the rent paid to the freeholder for the remainder of the lease term.

The method shown values the two interests separately and then determines the value of the leasehold interest by deducting the freehold value from the gross leasehold value.

It is suggested that in the valuation of any leasehold interest which involves an unusual set of letting terms, the opinion of value should at least be checked in this way.

7.8 CONTEMPORARY APPROACHES

Many investors and financial commentators have expressed impatience and frustration with the continuing use of the implicit 'non-growth' approach to valuation. It is claimed that not only does it misrepresent the property market and makes it difficult to know the effect of any yield adjustment but, and perhaps more important, it isolates property investment from other forms of investment, in that the calculated yields from each are not directly comparable.

It does seem unsatisfactory to assume the continuation into perpetuity of rent at a level that would be unacceptable, capitalized at a yield that grossly understates the level of return expected. That the fictions may compensate each other to determine the 'correct' value is considered less than satisfactory by many professional investors who are keenly interested in the rate at which rents are expected to rise and in the current and projected return on capital invested.

Starting in the early 1960s, several writers including Wood, Greaves (both for Ph.D. theses), Sykes, Brown, Fraser, Baum and Crosby have proposed various solutions to the problems inherent in the conventional approach but raised little interest outside a small group of mainly theoretical valuers. In 1976 Walls, an analyst with a firm of stockbrokers, criticized the way in which valuations were performed. The timing was particularly sensitive being shortly after the secondary banking collapse and arrangements for the financial 'lifeboat' encouraged by the Bank of England. He warned investors of what he considered to be shortcomings in the investment valuation method in a thin market and proposed a particular form of discounted cash flow approach. His precise approach was considered but not pursued although his views gained a good deal of publicity and support and generated a wider debate.

The RICS appointed a research fellow (Trott) to investigate the methods of valuation used in practice, resulting in two reports (1980, 1986). Trott's interim report suggested that Wood's 'real value' model was too complex for most practitioners to be able to use in their day-to-day work. It is debatable whether a valuer's ability and convenience should form criteria for the acceptance or otherwise of any particular approach although it is true that the presentation was obscure and gave the appearance of unreality. More recently, Wood's work has been remodelled and represented by Crosby in a form which he refers to as a 'real value/equated yield' hybrid and which is described in Chapter 8.

7.8.1 Discounted cash flow

The discounted cash flow (DCF) technique, a tool long used by accountants and business analysts, has been adapted for use in the appraisal of real property. The technique needed adaptation as it was now required to deal with very long time scales, whereas the business use of the method was normally for a short 'pay-back' period only, which could be as little as 5 or 10 years, with the terminal value most often relatively low (scrap value or nil).

The two approaches within the DCF technique are the net present value (NPV) and the internal rate of return (IRR). Net present value enables all the cash flows (incoming and outgoing) to be discounted at a selected rate of interest. When all the cash flows are summed (having regard to their signs) the result will show whether the target rate of interest will be achieved (0) or exceeded (a positive balance). A negative result shows that the particular discount rate specified will not be achieved.

Information is needed as to the rent payable, the current rental value, an estimate of the probable rental growth, the review intervals and the discounting rate of interest.

Valuers wedded to conventional approaches criticize the attempt to predict rental values and argue that the current rental value is as far as the valuer should go in providing information for a valuation. At the same time, they see no problem in using an all risks yield derived from property transactions even though it bears no direct relationship to other non-property yields; they are content to adjust that yield up or down to compensate for perceived differences in age, quality and earning capacity despite their inability to gauge the precise effect of any adjustment on actual return.

One of the advantages of the DCF technique is the discipline of quantifying the anticipated rate of rental growth which may cause the valuer to think more deeply about the qualities of the investment.

The importance of achieving rental growth may be judged from the knowledge that inflation has risen again recently and that the average annual rate of inflation in the 64 years from 1925 has been 5.2%.

The selection of an appropriate yield should cause less difficulty since a direct comparison may be made with other investment media. The surrogate for yield comparison has been the gilt-edged market. Long-term or undated gilts were regarded as the perfect risk-free investments, any other form of investment being inferior and with a level of risk, the difference being reflected in an addition to the rate of return.

The definition of risk has widened considerably in recent years to include in particular the problem of maintaining the purchasing power of income. Gilts today are more properly described as 'default-free'; some recent work, albeit on a limited scale, suggests that the inflation risk premium in gilts is itself of the order of 1.5%. It is questionable whether the return on gilts is any longer of much relevance to expected property yields given the conflicting nature of the two investments. But if it is to be used, any risk adjustment should arguably be made from the risk-free yield on gilts. In the case of high quality investment properties let to blue-chip tenants on modern leases with upward only rent reviews, it could be argued that there is no default risk and overall risk is significantly less than most investments in equities.

There is risk in holding any long-term investments when compared with short-term Treasury bills or similar investments: risk for which an investor will expect to be compensated by way of an increased yield. However, real returns are likely to remain fairly low given the continuing level of inflation and the quantity of funds, including international funds, seeking investments of quality.

Whereas the NPV assesses an investment relative to a selected yield, IRR determines the return produced by a particular income profit. The latter approach is therefore useful in non-market appraisals making use of the investor's required target rate or opportunity cost, as well as enabling an investor to compare two prospective investments.

Application of DCF techniques

The conventional valuation performed using an all risks yield is a dis-counting exercise, summing the value of the right to receive successive incomes for a stated period. The convention is to use a low yield to compensate for the expectation that income will rise, linked to use of the current rental value with no attempt to extrapolate the anticipated growth.

The discounted cash flow approach may be used in this way but its distinguishing feature is that it is more often used to reflect capital value using a market yield (derived from the overall investment market but possibly adjusted to reflect the peculiarities of property) combined with income in which anticipated growth is quantified. The following calcula-tions show the two approaches.

Example 7.17

An income of £1000 p.a. is received for a period of 3 years where the all risks yield is 5%, the market or equated yield 12% and the anticipated rental growth is 7%. It is assumed for the purposes of the example that the income may be increased each year.

By using the all risks yield approach (without explicit reflection of income growth):

Year	Income	PV £1 @ 5%	PV of income (£)	Cumulative total (£)
1	1000	0.95238	952.38	952.38
2	1000	0.90703	907.03	1859.41
3	1000	0.86384	863.84	2723.25

which gives the same result as

	£
Income	1000
YP 3 years @ 5%	2.72325
Capital value	2723.25

Where rental growth is to be taken into account, the valuer needs to estimate the growth and to use the appropriate discount rate when growth is included explicitly in the model (which is the market yield or opportu-nity cost).

Given the all risks initial yield and the appropriate market yield, the implied rate of growth may be calculated, as described in Chapter 8. Using the information given earlier, the valuation may be reformulated in the following way.

Year	Income with growth of 7%	PV of £1 @ 12%	PV of income (£)	Cumulative total (£)
1	1070.00	0.8929	945.30	945.30
2	1144.90	0.7972	912.71	1853.01
3	1225.04	0.7118	876.96	2729.97

Small error due to rounding.

The rental value is increased to the level anticipated at the end of each year whilst the discount rate is the opportunity cost, also known as the market or equated yield.

The opportunity cost is also referred to as the internal rate of return. The return in this case can be shown to be 12% in the following way.

Year	Outgoing (£)	Income (£)	PV @ 12%	PV of income (£)	Outstanding (£)
0	2729.97				2729.97
1		1070.00	0.8929	945.30	1784.67
2		1144.90	0.7972	912.71	876.96
3		1225.04	0.7118	876.96	0

The following example will demonstrate the use of the two main approaches of the DCF technique in the context of a valuation.

Example 7.18

Your client owns the freehold interest in well situated shop premises recently let on an institutional type lease at a rent of £25 000 p.a. for a term of 25 years with 5-yearly reviews. The all risks yield is found to be 5% from analysis of recent sales of comparable properties; the equated yield required is to be taken as 12%.

The valuation is straightforward: the shop is rack rented and there is market evidence of the all risks yield. The conventional approach would be

	£
Rent reserved being rack rent	25 000
YP in perpetuity @ 5%	20
Capital value	500 000

The NPV approach will now be used with an explicit reflection of the implied rental growth of 7.6355% p.a.

Years	Net rent (£)	Including growth at 7.6355%	YP 5 years @ 12%	PV £1 @ 12%	Capital values (£)
0–5	25 000	25 000	3.6048	1.00	90 120
6–10		36 117	3.6048	0.56743	73 877
11–15		52 214	3.6048	0.32197	60 602
16–20		75 434	3.6048	0.18270	49 681
21–25		108 979	3.6048	0.10367	40 727
26–∞*		157 443	3.6048	0.05882	185 215
					500 222

Small error due to rounding.
* Rent increases are not projected beyond year 26; the anticipated rent is capitalized in perpetuity at the all risks rate, thereby implying anticipated future increases in rent.

The capitalized amount is deferred at the equated yield rate. Where rental growth is known the equated yield can be found by taking two trial rates. The values are then calculated for both rates, one of which should give a positive result and the other a negative one. In that event, the true yield is captured between the two yields and may be determined more precisely by formula or by selecting two more yields and repeating the

Years	Rent (£)	at 11%			at 13%		
		YP	PV	£	YP	PV	£
0–5	25 000	3.6959	1.00	92 398	3.5172	1.00	87 930
6–10	36 117	3.6959	0.59345	79 217	3.5172	0.54276	68 947
11–15	52 214	3.6959	0.35218	67 963	3.5172	0.29459	54 101
16–20	75 434	3.6959	0.20900	58 268	3.5172	0.15990	42 424
21–25	108 979	3.6959	0.12403	49 956	3.5172	0.08678	33 263
26–∞	157 443	20.00	0.07361	231 788	20	0.04710	148 311
				579 590			434 976
	less purchase price			500 000			500 000
				+79 590			−65 024

calculation until the 'correct' yield is found when the sum of the discounted incomes is equal to '0'.

Taking the information given in Example 7.18 and using yields of 11 and 13%, the trial to find the IRR proceeds as follows.

The return is found to be somewhere between 11% and 13%. The yield may be found more precisely by using the formula

$$R_1 + (R_2 - R_1)\frac{NPV_1}{NPV_1 + NPV_2}$$

where R_1 is the lower discount rate used and NPV_1 is the NPV found from using the lower rate

$$= 11 + 2 \times \frac{79\,590}{144\,614} = 12\%$$

The result may also be demonstrated graphically by using the rules relating to similar triangles.

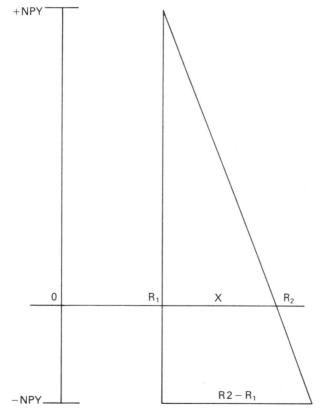

Then $\dfrac{X}{+\,NPV} = \dfrac{R_2 - R_1}{NPV + (-NPV)}$ and yield $= R_1 + X$

In the foregoing example

$$\frac{X}{79\,590} = \frac{2}{144\,614}$$

$$X = \frac{159\,180}{144\,614} = 1.100$$

$$R_1 + X = 11 + 1.100 = 12.100\%$$

Example 7.19

Some years ago the freeholder of shop premises granted a long lease (now due to expire in 25 years time) with irregular reviews of which only one, due in 10 years time, now remains. The current rent is £4000 p.a. whilst the current rental value is £10 000 p.a. The appropriate all risks yield deduced from transactions involving properties let on modern terms is 5.5%. The equated yield may be taken at 13% (with an implied rental growth of 7.28%). Assess the value of the freehold investment.

Years	Rent (£)	Growth @ 7.28%	YP @ 13%	PV @ 13%	£
0–10	4000	4 000	5.42624	1	21 705
11–25	(10 000)	20 192	7.32999	0.29459	43 601
26–∞		57 939	18.182*	0.047102	49 619
			Capital value of freehold interest		114 925

* At 5.5%.

Commentary

An important feature of any investment in property is the opportunity to renegotiate the rent at regular intervals and thereby maintain the purchasing power of the income. The opportunity is marred in this example by the timing of the reviews in the current lease. It is assumed that the property will be let on modern terms at the end of the current lease.

Relevant issues | 8

8.1 INTRODUCTION

Valuation practice may have changed little in recent years but there have been may articles, proposals and discussions as to the way in which the conventional approach should be applied or the newer approaches substituted. Meanwhile, much has changed in the world of property investment. The size of transaction may be much larger, the investment field is now world wide and the effect of inflation or growth is now a decisive factor.

There has been a preoccupation with dual rate leasehold valuations, their adjustment for inbuilt mathematical errors and taxation treatment, much of which is now seen as marginal, given the specialized and small market in which these transactions take place.

Many valuers regard much of the discussion on valuation methods as intrusive and lacking in market knowledge and experience and there is a strong body of opinion opposed to any attempt to look at new methods, particularly those requiring the valuer to make some prediction of rental growth.

No doubt valuers will continue to use that approach to valuation which they judge to be most likely to produce the 'right' result. But it is necessary, at least, that valuers understand and consider matters relating to such topics as risk and depreciation which need to find some expression in valuation practice. It seems likely that aspects of rental growth, inflation and equated yields will continue to be the subject of exploration and they are of course at the heart of the explicit approach. The important requirement is to provide the information required by the investor client whilst maintaining the integrity of the valuation process. The remainder of this chapter introduces some of these topics.

8.2 RENTS PAYABLE AT INTERVALS OF LESS THAN 1 YEAR

Printed valuation tables are unable to do more than give a range of values for each table, based on yearly intervals and those rates of interest most likely to be required by users of the tables. The way in which the tables are constructed assumes that the rent is not available for reinvestment until the end of the year.

Most modern leases provide for rent to be payable quarterly in advance with punitive surcharges for failing to do so. Rent in arrears is more likely to be paid at quarterly than yearly intervals. As a result, calculations of present value ignore the fact that some parts of the income may be available for earlier investment. Where the rent is paid quarterly in advance, the nominal yield used in the valuation tables understates the effective return.

Example 8.1

Assume a rent of £1000 p.a. payable annually in arrears, the market yield being 8%.

The capital value is

	£
Rent received	1000
YP perpetuity @ 8%	12.5
Capital value	12 500

in other words the rent of £1000 represents 8% return on the amount invested.

Where the rent is received not yearly but quarterly in arrear, the effective return allowing for reinvestment is 8.24% calculated as follows

$$250 \times \frac{(1 + j)^m - 1}{j}$$
$$\text{(one quarter's rent)}$$

giving annual income of

$$250 \times 4.12161 = £1030.40$$

or a percentage return of

$$\frac{1030.40}{12\,500} \times 100 = 8.24\%$$

where m represents the number of payment intervals in the year, i the annual interest payment and j the annual interest rate divided by the number of payment intervals).

Where rent is payable quarterly in advance the benefit is correspondingly greater. The monetary return is

$$250\left[\left(\frac{(1 + j)^{m+1} - 1}{j}\right)\right]$$
$$= 250 \times 4.20404 = £1051.01$$

or a percentage return of 8.41%.

The figures of years' purchase may be calculated direct from formula for payments received quarterly in arrears

$$YP = \frac{1}{4[\sqrt[4]{1 + i} - 1]} = 12.87$$

and for payments received quarterly in advance

$$YP = \frac{1}{4\left[1 - \sqrt[4]{\dfrac{1}{1 + i}}\right]} = 13.12$$

In each case the result is the year's purchase which the purchaser could afford to pay, and still receive an effective return of 8%.

Using the earlier example, the value would become

$$\text{Quarterly in arrears} = \text{Rent} \times \text{YP}$$
$$£1000 \text{ p.a.} \times 12.87 = £12\,870$$

and

$$\text{Quarterly in advance} = \text{Rent} \times \text{YP}$$
$$£1000 \text{ p.a.} \times 13.12 = £13\,120$$

An essential proviso is that the greater frequency of payment should be used only where the analysis has been carried out on a similar basis.

8.3 EQUIVALENT YIELD CALCULATIONS

Reversionary interests may be valued using the equivalent yield method or the term and reversion method. In the latter case, a downward adjustment is made to the market yield used in the capitalization of the term income to reflect the greater degree of security associated with it. Where such a differentiation is made, neither yield represents the overall non-growth yield. The equivalent yield will lie between the two rates used and may be calculated where required.

Example 8.2

Freehold warehouse premises are let at £5000 p.a. on a lease which expires in 6 years time. The current net rental value is £15 000 p.a. The appropriate market yield is 10%. What is the equivalent yield?

		£
Term rent	5000	
YP 6 years @ 8%	4.623	23 114
Reversion to	15 000	
YP perpetuity def. 6 years @ 10%	5.645	84 671
Capital value		107 785

A first approximation of the equivalent yield may be made by multiplying the difference between the yields by the fraction resulting from dividing the value of the term by the combined capital value and deducting the result from the higher yield, giving a value in this case of 9.79%. Where greater accuracy is required it is convenient to use a discounted cash flow model which will enable any remainder to be allocated.

Years	Rent/ capital value (£)	9%		10%	
		YP/PV	Amount (£)	YP/PV	Amount (£)
1–6	5 000	4.4859	22430	4.3553	21 776
6	150 000	0.59627	89 440	0.56447	84 671
			111 870		106 447
		deduct price	107 785		107 785
			4 085		−1 338

The equivalent yield (the IRR without growth) is calculated by linear interpolation

$$9\% + \frac{4085}{4085 + 1338} = 9\% + \frac{4085}{5423}$$
$$= 9.75\%$$

8.4 INFLATION

Before inflation was regarded as significant in investment terms, there was a differential between the yield expected from gilt-edged government stock and the return looked for from a good quality property investment; for long periods, the former was 3% whilst investors in property expected yields in the region of 5%. This differential — the yield gap — was accepted as a proper reflection of the various qualities of the two groups of investment, government stock being selected as a reference point because it was regarded as the perfect investment and risk free. In the early 1960s the effects of inflation were recognized. The emphasis on risk has shifted until now one of the major risks is the threat to purchasing power. The higher yield demanded is regarded as compensation for such risk but the stocks are now spoken of as being default free rather than risk free, reflecting the backing of the Government for the stock which however remains vulnerable to inflation, the future level of which is unknown.

Recent limited research has suggested that government stock itself carries a 1.5% risk premium. If this is so, the difference between the return on government stock and property investment returns should perhaps be less than previously thought.

Meanwhile, inflation increased the attraction of those investments where the income would tend to increase in line thus maintaining purchasing power. This happened at the expense of fixed interest investments. Government stock, typically carrying a nominal yield of 2.5% or 3%, began to sell at prices significantly below par value so as to show a much higher income return to the purchaser. On the other hand, low initial yields were acceptable on property investments since the rent was expected to grow and make the real return significantly higher over the life of the investment. At this point it became impossible to make a direct comparison between the returns on fixed interest and other investments, the difference between the two becoming known as the reverse yield gap — compensation for the inability of the return to keep up with inflation.

The loss of a direct reference point in gilt-edged securities meant that the valuer could no longer carry out a simple comparison of returns. In the ensuing years, various writers, mainly academics, have sought other methods of analysis to restore the ability to make a direct comparison. The favoured method has been some form of discounted cash flow using the market's typical target yield. The current opportunity cost of money is not a good guide to the long-term yield since the short-term rate may fluctuate over quite a wide range. In the same way, any view on the rate of rental growth must have regard for the long term and the ability to maintain the selected level year after year.

There is much unease in the profession about the use of the discounted cash flow approach though it is more readily accepted that it has a role in analysing an investment for a particular purchaser and has the ability to take account of unusual rent review patterns or expenditure. A common claim is that the valuer should not be taking part in rental growth forecasts, more properly the domain of the economist. This criticism appears to overlook the fact that any all risks yield is adopting a stance towards growth which can be exposed given information about the market yield requirement.

It is suggested that greater use of modern information technology and the need to respond to the analytical requirements of institutional clients will eventually result in explicit growth assumptions becoming an integral part of most investment valuations. Chapter 7 gives some examples of the approach.

8.5 INFLATION AND VALUATION

Crosby has made a detailed study of the valuation problems associated with inflation.

Noting the failure of Wood's attempt to change attitudes and acknowledging the innate conservatism of the valuation profession, he undertook a new study.

He describes his approach as a 'real value/equated yield hybrid' which looks at an income profile in terms of its purchasing power. He compares the end of year value of an inflation proofed income with one where an equated yield is used and a growth factor applied to the rent. He argues that both should give the same result and that therefore

$$\frac{1}{1 + i} = \frac{1 + g}{1 + e}$$

which is then rearranged in terms of e so that

$$e = (1 + g)(1 + i) - 1$$

where   i = all risks yield (decimal fraction)
       e = equate yield (decimal fraction)
       g = rent growth rate (decimal fraction)

He then investigates how the two methods would treat a fixed income and concludes that where the purchasing power declines and the income cannot be increased to compensate for that decline, it is necessary for the yield to reflect not only a real return but an additional return to reflect decreasing purchasing power. He goes on to suggest that, given the generally accepted link between the return on gilts and the return to be expected of property, allied to the analysis of initial yield to determine expected rental growth, an inflation risk free yield (IRFY) can be determined by rearranging the formula for e above so that

$$i = (1 + e)/(1 + g) - 1$$

Asserting that a fixed income declines in real value by the rate of inflation/growth until a review, when the purchasing power of the income is restored to its original value, he proceeds to apply his method to a situation where the initial rent is subject to review at 5-yearly intervals. Three figures of years' purchase are identified and may be used to value either rack rented or reversionary interests.

The underlying principle is that initial yields show an implied growth and that this information can be used to find a target rate of return, a rate that will be influenced by any change in the review interval.

Crosby has undoubtedly performed a service to the profession by his detailed study which should have the effect of highlighting the interaction of the components of any investment valuation. Whether his detailed and elegant mathematical exposition will encourage more ordered thinking, or even become generally adopted, remains to be seen.

The following examples should clarify the process.

Example 8.3

Four examples will be given to demonstrate the application of Crosby's real value/equated yield hybrid.

The property comprises shop premises in a good trading position worth a rent of £75 000 p.a. on full repairing and insuring terms, subject to review intervals of 5 years. The all risks yield is 5% and the equated yield 12%.

Four valuations are to be made.

1. A letting at the full rental value of £75 000 p.a.
2. A letting at £63 000 p.a. with a review in 2 years time.
3. A letting at £66 750 p.a. on a 3-yearly review pattern and a review in 2 years time.
4. A letting at £56 000 p.a. granted 5 years ago for a term of 17 years without review.

Before carrying out the valuations, calculations based on the all risks and equated yields need to be made to determine the implied rental growth and the inflation risk-free yield.

where　k = capitalization rate or all risks yield
　　　　p = implied rental growth over review period
　　　　g = annual rental growth
　　　　e = equated yield
　　　　t = review interval

First, calculate the implied rental growth, p
where capitalization rate k

$$= e - (SF \times p)$$
$$= e - \left[\left(\frac{e}{(1 + e)^t - 1}\right) \times p\right]$$

substituting

$$0.05 = 0.12 - \left[\left(\frac{0.12}{1.12^5 - 1}\right) \times p\right]$$

$$0.05 = 0.12 - 0.15741p$$

$$0.15741p = 0.07$$

$$\text{and } p = \frac{0.07}{0.15471}$$

$$= 0.44469 \quad \text{or} \quad 44.469\% \text{ over 5 years}$$

$$1 + p = (1 + g)^t$$

$$1.44469 = (1 + g)^5$$

$$(1.44469)^{1/5} = 1 + g$$

$$g = 7.635\% \text{ p.a.}$$

Now, calculate the inflation risk-free yield (IRFY)

$$\text{IRFY} = (1 + e)/(1 + g) - 1$$
$$= 1.12/1.07635 - 1$$
$$= 0.040554 = 4.0554\% \quad \text{(or 4.357\% with 3-yearly reviews and a value}$$
$$\text{for g of 7.324\%)}$$

Case 1
In a rack rented investment, the valuer would use the market evidence available and capitalize the current rental value at the appropriate all risks yield

	£
Estimated rental value	75 000
YP perpetuity @ 5%	20
Capital value	1 500 000

However, the '3 YP' calculation still holds good, producing the same result

	£
Estimated rental value	75 000

$$\text{YP} = \text{YP 5 years @ 12\%} \times \frac{\text{YP perpetuity @ 4.055\%}}{\text{YP 5 years @ 4.055\%}}$$

$$= 3.6048 \times \frac{24.6609}{4.4449} \qquad = \qquad 19.999$$

	£
Capital value	1 500 000

Case 2

Rent payable	63 000	
YP 2 years @ 12%	1.6901	106 476
Reversion to	75 000	

			£
YP perpetuity @ 4.055 % ×			
$\dfrac{\text{YP 5 years @ 12\%}}{\text{YP 5 years @ 4.055\%}}$	19.999		
		15.9431	
× PV £1 in 2 years @ 12%	0.7972		1 195 731
			1 302 204

Case 3

IFRY = 4.357% (see note above)

			£
Rent payable	66 750		
YP 2 years @ 12%	1.6901		106 476
Reversion to	75 000		
YP perpetuity @ 4.357% × $\dfrac{\text{YP 3 years @ 12\%}}{\text{YP 3 years @ 4.357\%}}$ = 20			
× PV £1 in 2 years @ 12%	0.7972	15.9438	1 195 791
			1 302 267

Case 4

			£
Rent payable	56 000		
YP 12 years @ 12%	6.1944		346 886
Reversion to	75 000		
YP perpetuity @ 4.055% ×			
$\dfrac{\text{YP 5 years 12\%}}{\text{YP 5 years 4.055\%}}$		19.9997	
× PV £1 in 12 years @ 4.055%	0.62065	12.4128	930 960
			1 277 846

8.6 NON-STANDARD REVIEW PATTERNS

With the emphasis on the need for regular rent reviews in any property investment for it to appeal to the general market, it is not surprising that there is an argument in support of adjusting the annual rent where the review interval departs from the normal 5-year period.

It is argued that the lack of reviews will affect the return available and lower the value of any property where this happens. The converse of the argument is that a tenant who is protected by the format of his lease from reviews at 5-yearly intervals should expect to pay a higher initial rent in compensation.

Tables of 'constant rent' and formulae are available to take account of particular patterns of review or lack of review.

The general formula (devised by Rose) is

$$R \times K$$

where R = the appropriate rental value where the property is let subject to 'normal' review intervals

K = the constant rent to be charged throughout the lease where there is no provision for review

$$K = \frac{A - B}{A - 1} \times \frac{C - 1}{C - D}$$

where

A = the amount of £1 for n years at the market yield

B = the amount of £1 for n years at the expected annual growth rate

C = the amount of £1 for s years at the market yield

D = the amount of £1 for s years at the expected annual growth rate

and   n = the length of the period during which no revision of the rent may take place

s = the normal rent review interval

Example 8.4

Calculate the rent level required to compensate a landlord who has granted a lease under an option to renew for a term of 21 years without any provision for rent review. The normal review interval is 5 years and the property would have let at a rent of £25 000 p.a. on this basis. The market yield may be taken at 12% and the anticipated rental growth at 5% p.a.

$$K = \frac{10.8038 - 2.9253}{10.8038 - 1} \times \frac{1.7623 - 1}{1.7623 - 1.2763}$$
$$= 1.26049$$

and rent to be charged (R × K) = £25 000 × 1.26049 = £31 512

The result suggests that where a lease is granted for 21 years at a fixed rent without review, that rent should be £31 512 where the appropriate rent on a lease subject to a 5-year pattern is £25 000 p.a. Whilst the figure is mathematically correct, it does mean that the tenant would be paying substantially more than what would normally be regarded as market rent in the first few years of the lease. The higher rent would place any business where the rent was a significant part of total outgoings in an unenviable position with respect to its competitors. Uncertainty regarding the growth of rents would contribute to the difficulty of negotiating a rent fair to and acceptable by both parties. Some valuers suggest that the lack of or longer review interval may be remedied by adding 1% to the rental value for each year the interval extends beyond what is regarded as normal. A further approach is to compare capital values and deduce a rent from the information.

Example 8.5

Taking the information in Example 8.4 it may be assumed that on a standard modern lease where the rent is to be reviewed every 5 years, the all risks yield by comparison with similar properties is 6.5%. A money market yield — the level at which a lease without review might be capitalized — is say 10%.

Valuation of property subject to modern lease terms

	£
Rental value	25 000
YP perpetuity @ 6.5%	15.38
	384 615

To maintain the same capital value where a higher capitalization rate is used, the rent will need to be greater. At a yield of 10%, the rent would be £38 462 (capital value/YP).

The capital value of the premises let on modern lease terms is

	£
Rental value	25 000
YP perpetuity @ 6.5%	15.38
Capital value	384 615

The rental value where there is no review becomes

	£
Rental value	x
YP perpetuity @ 10%	10
Capital value	384 615

	£
and x =	38 462

(an increase of over 50%)

Commentary

There is no doubt that a lease without reviews is less attractive to the vendor than one where there are regular reviews. But it is unlikely that the parties would be able to agree about the prospects of long-term rental growth.

The parties are moving from market evidence in the case of rents subject to review, to mathematical manipulation where there is no evidence of sales of property subject to long leases without review.

Even if a tenant was able to pay a rent higher than market rent in the early stages of his lease, would it be wise for him to do so, given the disadvantage he would suffer from that additional overhead in his trading?

The income is only partly secured on the property; the fact that it is let at a higher rent than would be sustained by the property in more normal circumstances would tend to make the investment more speculative and increase the yield demanded by investors.

There have been reports of investors obtaining an additional 1% rent for each year by which the review interval exceeds 5 years. This rule of thumb does not support the increase calculated in the above example.

8.7 IMPLIED RENTAL GROWTH

Low initial yields are acceptable to investors in property only because there is an expectation of rental growth. The expectation is not quantified; the all risks yield is derived from the market and represents its interpretation of the prospects. Nevertheless, the implications for growth can be exposed, once the expected market yield is known.

The market yield applicable to property will relate closely to the market yield of other, non-property investments once adjustments have been made to reflect differences in risk and other considerations.

The base normally adopted is fixed interest undated government stock to which a premium of 2% is added to reflect the additional risk inherent in investing in less than a perfect medium. The risk on government stock is almost entirely related to loss of purchasing power. Given recent research assessing the risk on government stock at 1.5% and reflecting on the relatively high average level of inflation of 5.2% p.a. over the last 64 years, it may be that the premium should be reassessed.

In some respects, investment in property occupied by major commercial or industrial concerns is likely, in its own way, to be as risk free as is investment in government stock. The opportunity to renegotiate the investment return at regular intervals from a guaranteed base in the shape of an upward only review provision is unique to property investment and an extremely valuable attribute.

As an individual property becomes older or where it is unlikely to attract users from the first rank of public companies, so the risk premium assumes more importance.

The principal point is that investment in property is not necessarily an inferior form of investment.

The growth rate implicit in the all risks yield can be calculated once certain information is known and assumptions made.

Example 8.6

An investor has the opportunity to purchase 2.5% consolidated government stock (undated) currently showing a yield of 9.8%. The investor is indifferent as to whether he buys stocks or invests in a tenanted property to show a yield of 12% taking account of anticipated rental growth.

Commentary
The annual rental growth rate assumed in this valuation may be calculated from the formula

$$\text{Growth rate p.a.} = \left[\sqrt{1 + \left(\frac{\frac{E - i}{E}}{(1 + E)^n - 1} \right)^{1/n}} \right] - 1$$

where   n = review interval
         E = yield reflecting rental growth (decimal fraction)
         i = all risks yield (decimal fraction)
        SF = annual sinking fund payment for review period at E%

Example 8.7

Assume a modern shop property let on FR and I terms with rent reviews at 5-yearly intervals which is available at a price to show an all risks yield of 5.5%. Substituting the known variables in the formula

$$\left[\sqrt{1 + \left(\frac{\frac{0.12 - 0.055}{0.12}}{(1.12)^5 - 1} \right)^{1/5}} \right] - 1 = 0.07158$$

An initial estimate may be made by deducting i from E (12 − 5.5 = 6.5%) but this result would be correct only where the lease was subject to annual rent reviews. The effect of adjusting the rent at 5-yearly intervals is to require a higher annual growth, in this case 7.16%.

8.8 ACCURACY IN VALUATION

The valuer aims for accuracy in valuation although it is well accepted that valuation is based on interpretation and opinion, and the view of one competent valuer is not necessarily the same as that of another equally competent valuer. Many valuers believe that they can value to within 5%; giving a range of 10%.

The courts have considered the question from time to time as in a case where Megarry found that he had to deal with the matter generally rather than with exact mathematics and he had to bear in mind that valuation was an art rather than a science (*Violet Yorke Ltd* v. *Property Holding and Investment Trust Ltd.*)

The Privy Council voiced their strong disapproval of any attempt to look at a valuation in too critical a fashion.

'In general their Lordships consider that it would be a disservice to the law and to litigants to encourage forensic attacks on valuations by experts where those attacks are based on textual criticisms more appropriate to the measured analysis of fiscal legislation.' (*A. Hudson Property Ltd* v. *Legal and General Life of Australia Ltd.*)

The courts have defended the need for a valuer to make assumptions.

'Often beyond certain well-founded facts so many imponderables confront the valuer that he is obliged to proceed on the basis of assumptions.' (*Singer & Friedlander Ltd* v. *John D. Wood and Co.*)

In that case (which related to development land where the propensity for error may be greater) it was accepted by both sides that the permissible margin of error was generally 10% either side of a figure which can be said to be the 'right' figure whilst in exceptional circumstances that margin may extend to 15%.

An experiment that drew much criticism was conducted by Hager and Lord and reported in a joint paper delivered to a meeting of the Institute of Actuaries. The authors stated their understanding from informal discussions that the range of valuations for any particular property would be about 5% either side of the average value. In seeking to test this hypothesis they invited ten surveyors 'who all have experience in asset valuation for pension funds, but do not necessarily have intimate knowledge of the locations chosen' each to value two properties and were very surprised to find a much greater range (summarized in Table 8.1).

Recent examples of wide disparities between the valuations of eminent firms of valuers has drawn much adverse comment on the valuation process. In the Oldham Estate incident, one firm prepared a balanced sheet valuation of £581 000 000 whilst another firm produced a valuation for the bidder of £436 000 000. The RICS acted quickly to investigate the valuations, but their claim that the valuations were undertaken within accepted professional standards was hardly reassuring given the Institution's unwillingness to elaborate on its statement.

Table 8.1 Summary of valuations obtained by Hager and Lord

Property A £'000	Property B £'000
630	450
653	525
680	550
710	600
712	Control ⎯⎯⎯ £605.000
Control ⎯⎯⎯ £725.000	610
745	615(2)
760	635
771	651
780(2)	655

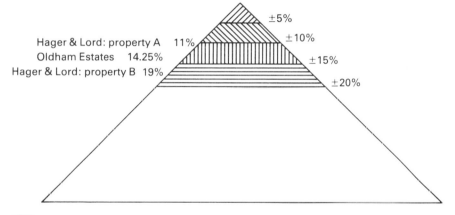

Hager & Lord: property A 11%
Oldham Estates 14.25%
Hager & Lord: property B 19%

±5%
±10%
±15%
±20%

Well within acceptable limits

Broadly acceptable

Acceptable in exceptional circumstances

Unacceptable

Figure 8.1 Range of acceptable error in valuation.

Peter de Savary took a whole page advertisement in *The Times* to give his account of the valuation of land on which he had received a loan from Blue Arrow and where there were alleged to be wide divergences of opinion, although not all the valuation information was made available.

The range of acceptable error is summarized in Figure 8.1.

A further factor leading to a differential between valuation and price is the 'weight of money' theory that suggests that institutions with a surplus of cash holdings may under certain circumstances be so anxious to invest that money in non-cash holdings that a higher price is paid in competition than the buyer's assessment of the intrinsic worth.

8.9 DEPRECIATION AND OBSOLESCENCE

Depreciation is particularly important in buildings which are intended to have a long life. It is defined as:

'the measure of the wearing out, consumption, or other reduction in the useful economic life of a fixed asset whether arising from use, effluxion of time or obsolescence through technological or market changes.'

The majority of valuers have been content to ignore the consequences of obsolescence unless the building to be valued is ripe for redevelopment within the next few years. Even an old building therefore is likely to be valued as producing a rent into perpetuity (where the land is of freehold tenure), the only quantitative reference to obsolescence being in the yield adopted in capitalization (and probably a fairly modest estimate of rental value).

A recent major research report by Salway identified the increased concern with obsolescence as an inevitable consequence of rapid techno-logical change. It called for changes in building design and a review of property management systems to enable identification of the optimal point at which to act.

The report is valuable in that it details the results of a study showing the discrepancy between the rental value of a modern office building and one 20 years old and concluding that when the higher yields required by investors are taken into account the older building will be worth little more than one-third of its newer counterpart.

The survey information shows the 25-year-old unimproved high street shop to be relatively unaffected by age when offered to let in contrast to other buildings of the same age. The difference is no doubt attributable to the extreme sensitivity of location so far as shops are concerned and the

Table 8.2 Annual rates of relative depreciation in rental value

Years	Offices (% p.a.)	Industrial (% p.a.)
0–5	3.3	3.1
5–10	3.4	3.9
10–20	2.7	3.2
0–20	3.0	3.3

Table 8.3 Percentage of locations experiencing the highest rate of fall-off in capitalization rate in the specified age bracket

Years	Offices (%)	Industrial (%)
0–5	12	48
5–10	72	44
10–20	16	8

high value of the site relative to the cost of replicating the building on the site.

The author investigated the rental value of offices and industrial buildings at 5-year intervals for the first 20 years in relation to a new building by using an equation first proposed by Brandon and Ward:

$$R = R(1 + d)$$

Where R = rental value of building n years old
     R = rental value of building o years old
     d = annual rate of depreciation in rental value

Using this equation the results shown in Table 8.2 were obtained.

The report also looked at the effect of age on capitalization rates, contributing to a significant depreciation in value (Table 8.3). When taken together with the increase in required yields to compensate for age, the difference is substantial.

Investors indicated a concern that yield levels when the report was prepared (1986), especially for offices and covered shopping centres, were too low to make adequate allowance for the depreciation of buildings. There is no doubt more to the problem than to ask why those investors expressing concern do not act by reducing their offers for such developments

The report suggests that the effect of physical deterioration of buildings can be controlled to some extent at the design stage by the use of life cycle costing techniques though it accepts that dealing with the problems of obsolescence are less straightforward. The recommendation asks for a great change in the attitude of developers who tend, quite naturally, to look for the least expensive initial solution without having too much regard for the implications on the life of the component or the cost of periodic maintenance.

Depreciation or failure during occupation is almost always a cost partly borne by the tenant by means of a full repairing lease or responsibility for maintenance through payment of a service charge. In a sense, therefore, maintenance work by the tenant subsidizes the developer's initial costs.

It is also suggested that the length of the lease and the circumstances of the particular development should be taken into account so that each lease expires at an appropriate time to enable development to take place.

The final recommendation of the report is that investors should avoid the purchase of overpriced assets and identify underpriced ones by means of a 'depreciation sensitive' appraisal technique described in the report.

Salway points out that in recent years yield changes have counter-balanced and partly hidden the effects of depreciation, enabling a satis-factory performance to be achieved even though the original price paid insufficient attention to expected depreciation.

Anyone reading the report must conclude that a 'prime' property remains in that category for a relatively short part of its economic life; it follows that purchasers with a preference towards prime investments will tend to acquire new buildings which are 'prime' in other respects also, and that there will be an active market to sell such properties when they no longer satisfy the definition.

8.10 PERFORMANCE

Property investment is made in the expectation that there will be a satisfactory return. Return is composed of annual income and anticipated capital appreciation. The provision for regular rent reviews and the effect on capital value of an increased rent have promoted property as a good hedge against inflation. Investors will periodically wish to check that their investments are performing as anticipated. The first measure will be the return on capital, represented by the relationship between income and original capital cost together with any later capital expenditure. In order to calculate the total return, the investor will also take into account the increase in capital value.

The formula for determining the total rate of return is

$$R = \frac{[C_1 - C_0] + [I]}{CE}$$

where R = total rate of return
C_0 = capital value at beginning of period
C_1 = capital value at end of period
CE = capital employed during period
I = income from rents during period

Example 8.8

An investment was purchased 5 years ago at a cost of £200 000. At that time and for each year since, the net annual rent has been £10 000. No capital expenditure has been incurred by the landlord. The rent has recently been reviewed under the terms of the lease and agreed at £13 382 p.a. for the next 5 years.

Calculate the total return from the investment on the assumption that: (1) the yield has remained unchanged at 5%; and (2) the yield required by investors has increased to 5.5%.

$$\text{(1)} \quad R = \frac{[267\,645 - 20\,000] + [5 \times 10\,000]}{200\,000}$$

$$= \frac{67\,645 + 50\,000}{200\,000} = 58.82\%$$

$$\text{(2)} \quad R = \frac{[243\,314 - 200\,000] + [5 \times 10\,000]}{200\,000} = 46.66\%$$

Commentary

Both results show the total return over a period of 5 years, but only where the income is received in one sum at the end of the period. Thus the return in the form of rent is understated since it will be payable normally in advance on a quarterly basis. The only way to obtain a true result is to calculate for each payment period, in this case quarterly. A capital value will need to be computed each time to reflect the influence on capital value of the shortening period to the next rent review. Each period rate of return is then aggregated to find the performance over the whole period.

Such calculations would be very time consuming by hand but are easily and quickly carried out with a computer using either a spreadsheet format or a customized program.

For more precision, time or money weighted rates of return may be used. The money weighted rate of return assumes investment and re-investment at a constant rate over the total period under review: the time weighted rate of return takes account of the lengths of the sub-periods.

8.11 MODERN AIDS AND APPROACHES

Opportunities exist to show that the residual method offers an approach at least as rigorous as other methods. The availability of computer facilities enables a residual model to be developed and tested in ways which it would be impracticable to attempt manually.

Freeing the valuer from the detailed and tedious process and associated calculations facilitates a higher level of understanding of the problem and possible solutions. A well conceived, constructed and tested program quickly undertakes the mundane calculations and eliminates the risk of arithmetical error. More importantly, it enables various statistical tests and observations to be made. The results of such tests may point to weaknesses within the approach whilst the very need to provide information acts as a check on unbridled enthusiasm. Both transformations are possible using current software.

Spreadsheets are particularly suitable to the calculations, enabling the valuer to pose a range of 'what if?' questions. The effect of a change in yield, rental value, building period or finance rate can be readily discovered.

Whilst the power and sophistication of spreadsheet applications for development appraisal may be of a very high order these are not obligatory characteristics. Processing power and sophistication are not always required to solve a particular problem. Even relatively simple spreadsheets offer significant assistance with calculations whilst being a source of considerable flexibility.

Successful exploitation of the potential depends upon a clear understanding of development appraisal problems and the capability of spreadsheets. The opportunity exists to run the program alongside accessible information banks and to incorporate the result in the word processing facilities.

The consequences of computer based-techniques are not all positive. The general tendency for using computer power to generate vast amounts of data simply because that facility is available applies equally to spreadsheet application. This should be resisted unless it can be shown that the user is fully conversant with the significance of the results.

Various statistical techniques may be employed to ensure that no undue risk is taken. The techniques include sensitivity analysis and probability.

8.11.1 Sensitivity analysis

Predictions as to the outcome of a development proposal rest upon the interaction of the various inputs. Those having the greatest influence on the final result are yield, rental value, building period and cost of finance. Others, such as fees and stamp duty, would have little effect on the final result even if they were to be increased very substantially. Simple programs offer the opportunity to change the significant variables in isolation or in combination, to show the effect on the residual land element or any other indicator chosen.

8.11.2 Probability

The degree of risk of significant change in the main variables is measured by assessing the probability of certain events occurring, assigning numerical values to the degree of probability thought to be associated with each particular variable.

Table 8.4 Probabilities of yield change

Yield of 5%	Percentage	Probability of change	Cumulative probability
+5% (5.25)	40	0.4	0.4
+10% (5.50)	30	0.3	0.7
+15% (5.75)	20	0.2	0.9
+20% (6.00)	10	0.1	1.0
	100	1.0	

It is possible to examine the probability of change in each variable and to calculate the combined probability. For example, it may be possible to assign probabilities to the likelihood of yield change in the period until the development is sold as shown in Table 8.4.

Similar estimations would be made for the other three variables. Even with a high degree of confidence in the forecasts, it is extremely unlikely that all would be as estimated.

The chance of all four main variables being correctly forecast, where individually each was given a 75% chance of being correct, would be less than 32%. Where a higher degree of confidence, say 90%, for each variable was achieved there would even so be a combined chance of less than 66%.

The residual method

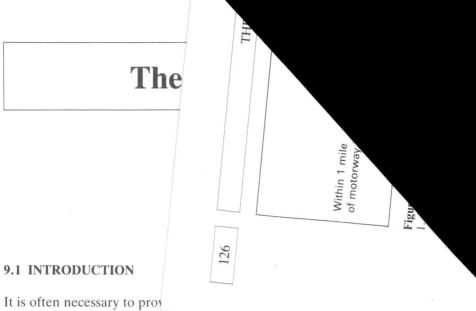

The

9.1 INTRODUCTION

It is often necessary to pro[...]
having obsolescent or otherwise unsuitable [...]
for redevelopment. The passing rent, even where there is one, relates to the current use and will not assist in determining the value of the land for redevelopment (though it may be one factor in deciding whether development or redevelopment is viable at a particular time).

In essence the problem appears to be a simple one; the market is likely to relate the value of the land to the level of profitability of the development. It may be that, in this quest for a value, recent sales of similar land for development will be of some assistance. For example, land for industrial development in any particular locality will tend to have a capital value within a known range; current sale prices would reflect supply and demand related to the physical advantages and disabilities of the particular site. The best price in that market would be obtained for level sites well served by a local road network and close to a motorway.

Where there has been a recent transaction involving a site similar in all respects to the one under consideration that is likely to prove the best evidence for the value of the site. Where the site differs in any material respect this will tend to be reflected in a change in the unit price. There may of course be other compensating considerations including competition between adjoining landowners in individual cases.

The value is often expressed as a capital sum per unit of measurement, i.e., acre or hectare. But the detail of the development, the size of the site and its access, the various expenditures involved and the potential for letting are likely to ensure that comparable information of the kind mentioned is not directly applicable and acts as no more than a rough check or perhaps as a useful shorthand way of reporting the result (Figure 9.1).

---

7 acres
of
Prime residential building land
in exclusive location
FOR SALE

Outline planning permission
for approximately 50 dwellings

junction

About 400 yards of frontage to unclassified road
Adjoining 18-hole golf course

---

e 9.1 Typical advertisement for sale of building land include:

Described as prime – gross density of approximately seven units to acre tends to confirm description. Will allow generous sites and large good quality dwellings (if demand exists);

2. Proximity to motorway junction should add to attractions of site to individual purchasers, especially where the owner visits various parts of the country on business;

3. The indication that the planning permission is for approximately 50 dwellings suggests that it is not firm and requires further enquiry;

4. The existing frontage should be able to be developed and will reduce development costs. The road is unclassified and it is therefore unlikely that there would be any highway objections to provision of individual access to each unit fronting the road;

5. The layout will no doubt benefit from the advantages of having a golf course adjoining the site.

Because of the scope for differences of significance, any attempt to determine market value by comparison pricing on a unit basis is likely to prove both difficult and unreliable.

9.2 THE BASIC APPROACH

The approach most often used is the residual method whereby all the costs of achieving the completed development are deducted from an estimate of the value of that completed development to arrive at the value of the site. Difficulties arise in the use of this method because the premises have not been built, rents are as yet not negotiated and plans and costings are tentative.

A market valuation is being placed on a completed development on the basis of information available on current rents and yields being achieved at the present time whereas the development may not be completed for 2 or 3 years or even longer, by which time the market may have undergone considerable changes, with a consequent effect on value.

Similarly, building costs or costs of short-term finance may be affected by external events, whilst any delay in the development schedule (often incurred due to complications in obtaining planning permission, interrup-

Gross development value	Rental value × YP giving Capital value on completion
less	
Costs of development	Building costs, planning application, fees, interest charges, etc.
Difference	Being the residual amount (needs adjustment to reflect purchase costs and holding costs, e.g. interest charges, for period of development)

Figure 9.2 Residual method in skeletal form.

tions due to inclement weather or an unsuccessful marketing campaign) is likely to have serious financial implications, especially where the cost of short-term finance is high.

So whilst no value profile should give a false impression of its degree of accuracy or certainty, it behoves the valuer to take great care to ensure that all data used are based on the highest quality of information available.

The great advantage of the approach is that it mimics the way in which the market considers the problem; in other words it is a market approach, shown in skeletal form in Figure 9.2. Considering Figure 9.2 it may be that the land is already owned by the developer or that the price has been agreed or is fixed. In such a case, a variation of the method may be used to determine the maximum amount available for building costs, the minimum net rent required form the completed works or the gross profit likely to result from the development. In each case the objective is the same; is a development of the site viable according to the criteria available (some, such as rents and yields, market derived; some stipulated by the particular developer).

9.2.1 The criticisms

The Lands Tribunal has commented many times on the use of the method. In the following extract from a decision published in 1965 it gives a reasoned and balanced account of the process.

'In the form in which it is normally presented to the tribunal, the residual method for the valuation of a development site shows a site value which is thrown up as the difference figure between the estimated value of a completed development of the subject land, and the estimated cost of carrying out that development. The figures of completed value and development cost are usually both much greater than the difference between them, i.e. they are both much greater than the site value which is being sought. In the make-up of both the completed value and the development cost there are a number of variables; the appropriate figure to be adopted for each of these variables will depend on the viewpoint as well as on the knowledge and experience of the valuer; the choice from which each such figure may be made is a fairly wide one, varying from what may be termed "conservative" to what may be termed "full"; and within this range whatever figure may be adopted is "correct" in the sense that it can be substantiated. From the viewpoint of a valuer who is retained by an intending vendor and who has therefore a responsibility to ensure that his client obtains not less than the full value of his land, there is a natural tendency to adopt somewhat full figures for the variables which together make up the completed value and/or to adopt somewhat conservative figures for the variables which together make up the development cost. Conversely, from the viewpoint of a valuer who is retained by an intending purchaser and who has therefore a responsibility to ensure that his client does not pay more than the full value of the land, there is a natural tendency to adopt somewhat conservative figures for the variables making up the completed value and/or somewhat full figures for the variables making up the development cost. At this point of divergence between the two valuers, however, the discipline of open market conditions intervenes, imposing external sanctions which are highly effective. The external sanction facing the valuer for the intending vendor is that, if his choice of figures for the variables should throw up too great a difference between completed valued and development cost, his client may well fail to find a purchaser at all because the calculated site value is above actual open market value. The external sanction facing the valuer for the intending purchaser is that, if his choice of figures for the variables should throw up too small a difference between completed value and development cost, his client may well fail to buy the land at all because the calculated site value is below actual open market value. It is a striking and unusual feature of a residual valuation that the validity of a site value arrived at by this method is dependent not so much on the accurate estimation of completed value and development cost, as on the achievement of a right balancing difference between these two. The achievement of this balance calls for delicate judgement but in open market conditions the fact that the residual method is (on the evidence) the one commonly or even usually used for the valuation of development sites, shows that it is potentially a precision valuation instrument. If there are two equally proficient valuers acting respectively for a willing vendor and a willing purchaser they would thus expect to agree on a price for the site in question, it being irrelevant for this purpose that one valuer may have arrived at the agreed site value by using figures for completed value and development cost differing substantially from those used by his colleague.'

It then goes on to consider the use of the method in cases where there is to be no market transaction.

'When a residual valuation is prepared for arbitration purposes, however, the conditions are very different; the valuation is then a calculation made *in vacuo*; and although there may be deemed open market there are no external sanctions acting as an incentive to the achievement of the delicate balance which I have described, because there is in effect a captive purchaser and a captive vendor. Thus there is no risk on the vendor's part of losing a sale by reason of the price advised by his expert being too high, nor is there any risk on the purchaser's part of missing a buy because the price advised by his expert is too low. Possibly as a side-effect of this absence of any external constraint, the natural tendency of the vendor's (or claimant's) valuer to adopt full figures when calculating developed value and conservative figures when calculating development cost almost invariably results (in the experience of the tribunal) in his putting forward an undependably high opinion of site value. Similarly the natural tendency of the purchaser's (or authority's) valuer to adopt conservative figures when calculating development cost almost always results in his putting forward an undependably low opinion of site value, and on occasion it may even throw up a minus site value. Having observed on so many occasions the working out of these tendencies in terms of widely conflicting "valuations" the deep impression on the minds of the tribunal is that under arbitration conditions . . . once valuers are let loose upon residual valuations, however honest the valuers and reasoned their arguments, they can prove almost anything (*First Garden City Ltd* v. *Letchworth Garden City Corporation* (Ref/74/1964). It is against this background and for this reason that the tribunal has reluctantly found itself compelled to reject the residual method when put forward as opinion evidence, unless there is no simpler method of valuation available'. (*Clinker & Ash Ltd* v. *Southern Gas Board*).

As will be seen, these remarks were made in a case where the Tribunal was required to settle a dispute over the amount of compensation payable where land was being acquired compulsorily. There was a strong opinion that a claim built up by use of the residual method was unsatisfactory except when exposed to the rigour and proof of the market place in a subsequent arm's length transaction.

Builders, developers and others wishing to develop land for a particular purpose will quite naturally move towards the site value by a deductive process — that is, they will ask themselves the potential value of the completed development, the costs of achieving that development and the return required for the effort, skill and risk involved in the transformation.

It should be noted that the difference indicates the maximum amount available for purchase by that particular developer rather than the value of the site; other developers may arrive at higher or lower amounts reflecting levels of efficiency, costs of overheads, the individual firm's cost of borrowing or the strength of the developer's desire for that particular piece of land. This last statement should be emphasized. In any assessment, the residual amount will reflect the type of development, its letting or sale value, the anticipated time to complete ready for occupation, the expected length and result of the marketing campaign and the cost of finance meanwhile. These and a number of other factors will influence the end result.

To take an example, an efficient and well-established builder or developer may be able both to borrow more cheaply (because of his reputation) and build more quickly than some of his competitors for the site. As a direct result, his calculation indicating the price he is able to pay will disclose a larger sum available for site purchase than that available on the basis of similar calculations made by others. In determining the amount he should offer, he should take into account not only the residual amount disclosed by his calculations but the likely competition, the current state of the market and the uncertainty associated with this particular transaction.

The unique calculations made by each interested party will result in a range of available bids. The greater efficiency of the builder in our example will ensure that he can offer more, although he should be able to make the winning bid without sacrificing the whole of the increment attributable to his greater efficiency or other particular circumstances.

It is possible that the end result will show a negative amount. In that case, the exercise suggests that, on the assumptions made, development of the site is not viable at the time of the calculation for the use and at the density envisaged. Time has not been wasted in this case: on the contrary, the exercise has served a purpose in indicating to the prospective developer that this particular site should be avoided unless there is a real prospect of reducing costs and or intensifying development to increase the value of the completed development. It is in this way that development proposals are shaped and refined.

9.2.2 Widespread use

The method is unique in that it is widely employed by both valuers and by developers and builders, i.e., 'the market'.

At its simplest, the procedure is little more than a mental calculation, a process which may be acceptable where the site is a small uncomplicated one for a single unit of development and the calculation is made by someone having wide experience in the particular type of development envisaged and able to relate to a recent similar site.

It would not be prudent to try to assess larger or more complicated sites in the same way particularly since the method has attracted much criticism due to its lack of sophistication. It is possible to identify the variables that are critical to the calculations and then to test for a range of values. For example, suppose the valuer identified five variables, any great alteration in which would have a significant effect on the result. He could then quantify a possible range representing the expected outcome, the worst outcome and the best outcome. To calculate this range of number in every combination would result in 3125 individual answers. It is apparent that a busy valuer would be unlikely to have the time to make such a set of calculations. However, there is no such difficulty with the aid of a computer.

The ability to test a range of inputs and then to apply statistical tests should make the method much more reliable and also more acceptable in those non-market situations where its use has previously been criticized.

It is proposed to describe the basic mechanics of the method before considering some of the refinements and checks available.

10	**The process**

Having collected as much information as possible, the valuer will make use of it in determining the residual value of the site. (It is worth repeating that the available information may be used to produce other results; for example, the land may already be in the ownership of the developer or the price already fixed; the developer may wish to determine how much is available to spend on construction works or the minimum value of the completed development necessary to justify the payment of a certain land price.)

Each item is considered in more detail. Table 10.1 lists the main variables to be considered.

10.1 VALUE ON COMPLETION

The first step is to estimate the capital value of the completed development. For this exercise, it will be necessary to calculate the lettable floor area and the unit letting value to find a total rental value to which is applied a yield derived both from analysis of the sale prices of other similar developments and experience of yields demanded by the market in general. The end valuation is normally straightforward, envisaging a rack rent capitalized in perpetuity or for the length of the ground lease available (a long period, 99 years or more probably 125 years unless the ground lease was granted some time ago when the residue will be available). It is likely that the market will reflect the nature of a leasehold tenure by the expectation of a slightly higher yield. Where a leasehold interest is concerned, the frequency and nature of rent reviews of the head leasehold interest will be relevant to yield as will the effect of any unusual provisions in the lease which fetter the developer's ability to manage and operate the development as he thinks best.

Table 10.1 Main variables

For gross development value
    rental value
    outgoings
    management
    ground rent
    all risks yield

For development costs
    demolition
    site clearance and preparation
    building cost (main contract)
    building cost inflation
    car park provision
    landscaping
    external signing
    contingencies
    off site works
    section 106 agreements
    fees of professional team
    local authority charges for planning application
    local authority charges for building regulation approval
    insurance

For fees on site acquisition
    agent's fees
    legal fees
    stamp duty

For fees and expenses on development
    surveyor's fees
    legal fees
    letting charges
    marketing costs
    compensation to existing tenants
    compensation/payment to occupiers of adjoining properties

For finance on site acquisition
    short-term finance until fully let

For finance on development
    short-term finance until fully let (instalments drawn down as required during
    progress of building works)

Value Added Tax will be payable as an input tax on costs, fees, etc. It may be recoverable
in whole or in part.

10.2 DEVELOPMENT COSTS

It may be that at the time of the initial investigation of financial viability
the developer has no more than a sketch design and that details of
construction problems, design and finish are incomplete in which case it

will not be possible to price accurately. The valuer should be concerned to earmark a sufficient sum for building costs and contingencies, especially since these have a 'knock-on' effect on other costs (for example, fees and finance). Too generous an allowance, on the other hand, may make the residual amount uncompetitive and result in failure to purchase the site. So a good deal of expertise and experience is called for at this stage.

The building costs are calculated on the gross area within the internal walls without any deduction for internal walls, stairwells, lift shafts and the like. The calculations should also include associated items such as demolition, site clearance and preparation where relevant.

10.3 PROFESSIONAL FEES

The fees charged by architects, quantity surveyors and engineers may account for between 12% and 15% of the construction cost depending on the complexity of the building. Local authorities now charge for planning application and building regulation approval.

Site acquisition, the development itself and any sale of the completed scheme will involve fees, charges and expenses. The developer may accept responsibility for the professional fees of other parties (the site vendor, the eventual purchaser of the development) and will be responsible for an arrangement fee in respect of the finance. In addition stamp duty and VAT will be payable. The total cost of fees and VAT is likely to be as much as 25 or 30% of the total costs of development.

10.4 BUILDING CONTRACT

The building contract may be negotiated subject to a 'rise and fall' clause — the usual method whereby the client is exposed to any future changes in wage rates and material costs occurring after the commencement of the contract. Alternatively, the contractor may be prepared to proceed on the basis of a 'fixed-price' contract which is likely to be at a higher figure since he is taking all the risks of future price increases. In estimating the cost under this approach, the contractor will have regard to the length of the contract, the type of labour required, his ability to reserve materials for later delivery at a fixed price, the state of his order book, the general level of activity in the industry and his view of the wider economic portents.

10.5 OTHER COSTS

It is common to include a 'contingencies' allowance of 5% or so to cover the cost of unexpected items — such as additional foundations — which may become apparent only once construction has commenced; this item will be subject to fees also.

10.6 SHORT-TERM FINANCE

Finance during the term of the development tends to be expensive with interest rates well above long-term interest rates. The level of interest rate will depend not only on minimum bank lending rate but on the size of the facility (in itself and as a proportion of the total costs), the track record of the borrower and whether the loan is part of a larger transaction (e.g., where the finance is provided at a concessionary rate by an institution or pension fund which has agreed to buy the investment once completed and let). In some cases the developer will have access to his own funds which will either eliminate or reduce his dependence on external borrowings.

All these considerations relate to individual circumstances and should not intrude on the present calculations for the purpose of which a market rate should be taken. Any benefits arising from the particular situation of the developer should then be taken into account in determining whether to offer more than the amount indicated by the residual amount. Should he decide to do so, he will not of course achieve the target represented by the calculations. On the other hand, use of the knowledge at the negotiating stage enables the developer to reflect the extent of his wish to acquire that particular site.

Any finance arranged will be required to purchase the site and to fund the building activities until the development is complete and all the units let. The borrowing for the land purchase will be at a constant level for the whole of the period, whereas the amount required to finance the building works will increase as the building progresses. In the absence of more adequate information, it has been the practice to assume that the gradual increase in borrowing over the period of construction can be equated with an average debt for the whole period of 50% of the total cost. This is of course no more than an approximation where conditions are normal. Where better information is available it should be used, especially where it is unlikely that the usual practice will fully represent the pattern of borrowing. The requirement for short-term finance is shown in Figure 10.1.

Once the building work has been completed, it is assumed that finance will continue to be necessary until the letting of the units has been finalized. Finance costs may be quoted on a quarterly or annual basis and the total cost 'rolled up' so as to become payable when the development is finally completed and let.

Where the interest is calculated on other than yearly rests, account should be taken of the actual costs involved. For example, on a finance

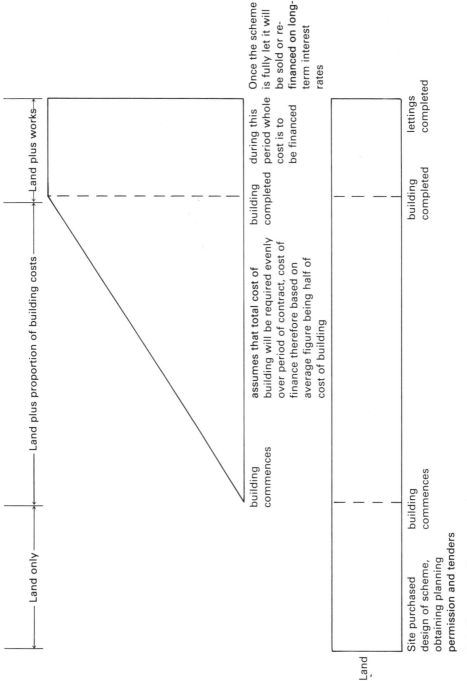

Figure 10.1 Short-term finance requirements.

Land only

Land plus proportion of building costs

Land plus works

Once the scheme is fully let it will be sold or re-financed on long-term interest rates

during this period whole cost is to be financed

assumes that total cost of building will be required evenly over period of contract, cost of finance therefore based on average figure being half of cost of building

building commences

building completed

Site purchased
design of scheme,
obtaining planning
permission and tenders

building
commences

building
completed

lettings
completed

Land

cost of 16% p.a., the true rate where calculated on quarterly rests is approximately 17%.

The following example shows the influence of finance costs and the increases occasioned by delay.

Example 10.1

The total development cost is £1 000 000 and the building period 12 months with a further 3 months to complete the lettings. The developer has negotiated a borrowing rate of 16% p.a.

The cost of finance is
Building period 1 year at 16% on average of half total cost

$$500\,000 \times 0.16 \times 1 \quad = \quad 80\,000 \text{ £}$$

Letting period 3 months at 16% on total cost

$$1\,000\,000 \times 0.16 \times 0.25 = \quad 40\,000$$

Total cost   120 000

Now assume that a bad winter delays building which is now expected to take 18 months, whilst a bleak economic forecast makes it likely that it will take 6 months to complete the letting programme. The cost of finance will increase appreciably.

Building period 18 months at 16% on average of half total cost

$$500\,000 \times (1.16 \uparrow 1.5) - 1 = 124\,679 \text{ £}$$

Letting period 6 months at 16% on total cost

$$1\,000\,000 \times (1.16 \uparrow 0.5) - 1 = \quad 77\,032$$

Total cost   201 711

increasing finance costs by approximately two-thirds.

None of these calculations includes the cost of financing the land purchase.

The finance cost would be appreciably higher where interest is calculated on quarterly rests.

Example 10.2

Recalculate interest payable on the development described in Example 10.1, except that interest is to be chargeable quarterly.
Where there are no delays and no difficulty in letting the units, the costs would be

                    £

Building costs £500 000 for 12 months 500 000 × (1.04 ↑ 4) − 1 = 84 929
Letting period £1 000 000 for 3 months
         1 000 000 × (1.04 ↑ 1) − 1 = 40 000

making a total cost of short-term finance of   124 929

The costs associated with longer building and letting periods are
Building costs £500 000 for 18 months 500 000 × (1.04 ↑ 6) − 1 = 132 660
Letting period £1 000 000 for 6 months
         1 000 000 × (1.04 ↑ 2) − 1 = 81 600

making a total cost of short-term finance of   214 260

10.7 ADVERTISING AND MARKETING

The cost of publicity in general can be very high and is not entirely predictable; where lettings are not taking place as expected, it may be necessary to relaunch the campaign and increase the budget by a significant amount. Expenditure may include newspaper, magazine, radio and television advertising, site boards, show building, negotiators on site, printing of brochures and other publicity material, press releases and general launch costs. The cost is best estimated by costing the proposed campaign and adding a margin; it would be inappropriate to calculate the cost as a percentage of the capital or rental value unless the developer had considerable experience with the particular type of development.

10.8 FEES AND COSTS

The cost of letting and sale (where a sale is to take place on completion of lettings or where competing developers are likely to operate on this basis) should be included. Such fees are likely to be related to the total value of the development. Where more than one agent is engaged, the total costs are likely to be higher. Costs may be incurred in compensating outgoing tenants or in winning the consent of occupiers of adjoining property (for example, infringements of rights of light or allowing a crane to traverse the air space). All letting and sales transactions will be liable to stamp duty.

10.9 DEVELOPER'S PROFIT AND RISK

It is usual for a developer to include a sum for profit and risk. The amount charged may be a percentage of the capital value of the completed development or a percentage of the total costs involved, the allowance typically ranging from 10 to 25%.

The level of profit and risk will be judged on the complexity of the proposal, the volatility of the outcome, the prestige of being associated with the particular development and the extent to which the profitability may be assured (possibly in part by a pre-arranged sale to a fund).

10.10 LAND VALUE

The gross amount available for land purchase may now be found by deducting the total estimated costs from the completed development value.

The remaining figure includes the costs of acquisition and finance incurred from the date of purchase until it becomes income producing on completion of the development.

The net amount after deduction of costs is not a value in the strict sense, merely an indication of the maximum price available for the site if the required returns are to be achieved on the basis of the information available.

Example 10.3

Using a short-term interest rate of 12% and a total holding period of 2.5 years, the gross residual value will include the 'rolled-up' finance cost plus the other costs and interest thereon. An example will show how the net development value is calculated where the gross residual land value is £650 000.

	£
Gross residual land value	650 000

where x = net residual land value the cost of
finance on land for 2.5 years at 12% is $0.32753x$
the acquisition cost plus finance for 2.5 years at 12% is $0.05310x$.
Total costs are therefore $0.38063x$
and $1.38063x$ = the gross value (£650 000)
Therefore x (the net residual land value) is £470 799
The calculation may be checked

Purchase price of land	470 799
Acquisition costs at 4%	18 832
Total costs	489 631
Interest at 12% for 2.5 years	132 753
Gross value including interest and costs	650 000

The importance of treating the gross residual value in this way cannot be over emphasized. In the fairly typical situation shown above, the net price available for land purchase is only approximately 72.5% of the gross value.

The following examples show the whole process at work; in each case the commentary expands particularly on aspects not previously considered in detail.

Example 10.4

A development company wish to acquire a site in a provincial city where outline planning permission exists for the erection of an office block having a gross floor area of 1500 m². The accommodation will be on four floors with basement parking for 20 vehicles. There is a steady demand for office accommodation at rents in the region of 120 per m². Short-term finance is available at 13% and the completed development could be sold to show a yield of 7%. Construction costs are likely to be in the region of £600 per m² and the building period 12 months. The company wish to know the maximum price it can afford to pay for the land.

		£	
Value of scheme on completion			
rental income 1350 m² @ £120 per m²		162 000	(1)
YP in perpetuity @ 7%		14.285	(2)
Capital value		2 314 286	(3)
Costs of development			
building costs @ £600 per m² (1500 m²)	900 000		(4)
contingencies @ 5%	45 000		(5)
professional fees @ 14%	132 300		(6)
finance costs @ 13% p.a.	105 037		(7)
	1 182 337		
marketing costs, say 20% of rental value	32 400		(8)
letting fees and sale fees	114 171		(9)
developer's risk and profit at 20% of completed value	462 857	1 791 765	(10)
Gross residue available		52 252	

deduct acquisition costs @ 4%
and interest during total holding
period at 13% p.a. Let x = net amount,   (11)
then finance for 15 months @ 13% = $0.1651x$
and acquisition costs of 4% for 15 months @ 13% = $0.1651x$
Thus $1.2153x = £522\,521$

	80 347	
Net amount available for site acquisition	429 952	(12)

Commentary

The example shows a basic calculation which will indicate whether a scheme is likely to prove viable, based on the assumptions made. It should be borne in mind that the initial study is very approximate; it will be noted that the planning permission exists only in outline. The valuer will be making many assumptions about the design and quality of the building which will affect rental values, yield and building costs. The outcome in terms of the value of the land will need to be treated with caution.

The following specific comments are made, relating to the numbers appearing in parentheses.

1. The gross floor area given has been reduced by 10% to arrive at the net area for letting purposes.
2. The yield is determined by reference to known comparables.
3. The capital value is the best estimate of the completed value of the development. There is an element of risk in the assessment since the building is not yet available and the market for office accommodation may change before leases are finalized.
4. Building costs are assessed as accurately as possible in the light of information currently available.
5. It is usual to make an allowance for costs incurred on items not anticipated when the building costs were assessed.
6. Professional fees include expenditure on the services of architect, quantity surveyor, mechanical and electrical engineers and others where required.
7. It is assumed that the developer will incur interest charges on building finance or that if he uses his company's funds there will be an opportunity cost. In the absence of more detailed information on the building programme and timetable, it is assumed that the amount borrowed over the building period will average half the building cost (including contingencies and fees). The short-term finance facility is likely to carry a fairly high interest rate. Where there is a possibility that the building will not be fully let on completion, finance will be required for a further period. At this stage, the whole of the building cost will be borrowed or assumed to be borrowed. In this example, where the building costs amount to £1 077 300, the building period is 12 months, the rate of interest is 13% and it is considered necessary to allow a further 3 months in which to achieve full lettings, the calculation of interest would be

$$(1\,077\,300 \times 0.5 \times 0.13) + (1\,077\,300 \times 0.25 \times 0.13) = £105\,037$$

The approach assumes that interest would be charged on yearly rests. It is more likely that the interest would be calculated at three monthly intervals although the cost would be 'rolled up' and the interest paid in one sum at the end of the project. The calculation would be made by increasing the period four-fold and dividing the interest by four. In the above example, the costs would increase as shown by the workings

$$1\,077\,300 \times 0.5 \times 0.0325^4 + 1\,077\,300 \times 0.0325 = £108\,525$$

8. and 9.   A proper marketing budget should be allocated to which should be added the estimated letting fees. The provision for sales fees is on the assumption that the completed development will be sold.
10. The developer is engaging in risk and is also wishing to make a profit on the operation. The combined provision will vary with the complexity of the development and possibly how anxious the developer is to be involved with the work. The allowance is normally expressed as a percentage of the completed value or of the total building costs.
11. The difference between the capital value and the costs of development is the maximum amount available for the purchase of the site. But it includes the financing costs of holding the land for the period up to full letting and for acquisition costs usually taken at 4% to allow for fees and stamp duty. The net amount available for purchase of the land is calculated as follows (x = net residual amount).

Finance for 15 months at 13%	$= 0.1651x$
Acquisition costs for 15 months @ 13%	$= 0.0502x$
$1.2153x = £552\,521$ and $x = £429\,952$	

12. The developer should not pay more than £471 800 for the land. Should he do so, he will not achieve the desired return unless savings are made elsewhere sufficient to offset the increased payment.

The further examples given will include notes only where additional explanations are necessary.

The next example looks at the effect of various changes in elements of cost. It is usual to apply the principles of sensitivity analysis to development appraisals. Some changes are more sensitive than others: a change in estimated rental income, yield, building costs or building period could have a significant effect on the result. Even simple computer programs are capable of providing information on the effect of any change in the data used.

Example 10.5

The facts are as given in Example 10.4 except that the land is being transferred by a subsidiary company at a price of £600 000. The development company wishes to know whether the proposal would show a profit.

The previous calculations are adjusted to take account of the additional information.

Value of scheme on completion £

rental income 1350 m² @ £120 per m² 162 000

YP in perpetuity @ 7% 14.285

Capital value 2 314 286

Costs of development

building costs @ £600 per m² (1500 m²)	900 000	
contingencies @ 5%	45 000	
professional fees @ 14%	132 300	
finance costs @ 13% p.a.	105 037	
	1 182 337	
marketing costs, say 20% of rental value	32 400	
letting fees and sale fees	114 171	
cost of land	600 000	
fees and stamp duty on transfer @ 2%	12 000	
finance for 15 months @ 13%	113 016	2 053 924
Balance being developer's risk and profit		260 362

Commentary

The calculations have been adjusted to take account of the information given regarding the price of the land.

The residue in this case is the amount available to reimburse the developer for risk and profit. It is equivalent to 13% on the value of the completed development (compared with 20% provided for in Example 10.4).

A cash flow approach enables the valuer to regard more precisely the timing of cash flows which may have a significant effect on viability, especially where short-term interest rates are high or development periods long.

The next example will demonstrate a variety of such techniques and show how information may be used more sensitively.

Example 10.6

Value a cleared site for a warehouse development for which planning permission exists. You have collected the following information:

	£	
	£	
Rental value	170 000	
All risks yield	10%	
Building costs	675 000	(payable quarterly, 15% 30%, 25% and 30%)

Professional fees	94 500	(75% initially, balance on completion)
Marketing expenses	25 000	
Letting fees	17 000	
Sale fees	46 750	
Building period	12 months	
Short-term interest rate	14% p.a. (quarterly rests)	
Developer's profit and risk	15% of completed capital value	

It is anticipated that units with a total rental value of £100 000 p.a. will be let and occupied immediately on completion of work and that the remaining units will be let by the end of the next quarter.

The information is first used in a traditional residual calculation and then three cash flow approaches are used. Whereas the traditional method relies on a good deal of approximation, the cash flow approaches give the opportunity to build in as much precision as required, bearing in mind that very little of the information used is certain and that price increases, economic crises or increased materials and labour costs may occur and affect the results. However, the more careful treatment of the information available removes some of the force of the criticism of this method.

The following cash flow techniques are shown in worked examples.

1. A cash flow approach by charging period.
    The period selected for demonstration purposes is one quarter; a developer monitoring the progress of a scheme would expect a major financial report at least once a month and access to the flow of funds on a daily basis. A modern computer program would provide such information routinely; in doing so, some of the risk of development is removed, since there is much more opportunity to review costs and be alerted to problems at an early date.
2. A discounted cash flow approach.
    In this case all expenditure is discounted and deducted from a similarly discounted capital value, reflecting the elapsed time before the investment may be realized. The advantage of the former approach is that it shows the actual indebtedness at the end of each charging period, rather than the discounted cost and is therefore more helpful for budgeting purposes.
3. A net terminal value approach.
    Expenditure in each charging period is debited with the total borrowing charges for the remainder of the development period and to that extent overstates the costs outstanding at any one time.

The example is a simple one but benefits from the cash flow technique. Where payments or receipts are irregular and particularly where the development consists of a number of units which may be let or sold at different times, the cash flow approach is the only way in which an accurate account of the proposal can be presented.

			£
Gross development value			
Rental value		170 000	
YP in perpetuity at 10%		10	
Capital value			1 700 000
Less costs of scheme			
Building costs	675 000		
Fees	94 500		
Short-term finance @ 14% pa	75 000		
Marketing expenses	25 000		
Letting fees	17 000		
Sale fees	46 750		
	933 250		
Developer's profit @ 15% of CV	255 000		1 188 250
Gross amount available for site purchase			511 750
PV £1 in 15 months @ 14%			0.8489
			434 437
Less costs of acquisition at 4%			16 709
Current site value			417 728

1. Cash flow approach by charging period

Period	Item	Outgoing (£)	Income (£)	Net cash flow (£)	Outstanding from previous period (£)	Interest at 3.5%	Total outstanding (£)
1	Costs	101 250					
	Fees	70 875		(172 125)	—	—	172 125
2	Costs	202 500					
	Marketing	12 500		(215 000)	172 125	6 024	393 149
3	Costs	168 750					
	Marketing	12 500		(181 250)	393 149	13 760	588 159
4	Costs	202 500					
	Fees	23 625					
	Letting fees	10 000	25 000	(211 125)	588 159	20 586	819 870
	Rent						
5	Letting fees	7 000					
	Sale fees	46 750					
	Rent		42 500	(11 250)	819 870	28 695	859 815

		£
Sale price		1 700 000
Less Total outstanding	859 815	
Developer's profit @ 15% of sale price	255 000	1 114 815
Gross amount available for site purchase		585 185
PV £1 in 15 months @ 3.5% per quarter		0.842
Current site value and acquisition costs		492 726
Less costs of acquisition @ 4%		18 951
Current site value		473 775

2. Discounted cash flow approach

Period	Item	Net cash flow (£)	PV @ 3.5% quarter	PV of cash flow
1	Cash flow	(172 125)	0.9662	(166 307)
2	Cash flow	(215 000)	0.9335	(200 703)
3	Cash flow	(181 250)	0.9019	(163 469)
4	Cash flow	(211 125)	0.8714	(183 974)
5	Cash flow	(11 250)	0.8419	(9 471)
	Developer's profit	(255 000)	0.8419	(214 685)
	Sale price	1 700 000	0.8419	1 431 230
	Current site value including costs			492 621

(from which the site value will be calculated as before)

3. Net terminal value approach

Period	Net cash flow (£)	Interest @ 3.5% per quarter	Total expenditure to completion (excluding land) (£)
1	(172 125)	1.1475	197 513
2	(215 000)	1.1087	238 371
3	(181 250)	1.0712	194 155
4	(211 125)	1.0350	218 514
5	(11 250)	1.0	11 250
		Total	859 803

This again enables the site value to be ascertained.

The profits method

PART FIVE

The profits principle | 11

Most commercial premises are occupied by tenants intent on trading at a profit; the level of anticipated profit will be one of the determinants used by the tenant in determining the maximum rent that he is able to pay. Whereas the level of profit expectation determines a trader's ability to pay the rent, the valuer does not usually have access to such information and relies on the interaction of the market as evidenced by the level of rents agreed.

11.1 RETAIL OUTLETS

In a town centre where retailers compete with one another, there will be sufficient leasing activity to provide market evidence which will indicate the optimum rental value of a particular unit. The share of gross profit available for payment of rent will vary from one trader to another and is likely to be influenced by the type of goods sold, the volume of sales and the rate of stock turnover, the mark-up and the space required. For example, a jeweller needs less space than, say, a butcher, but sells items which are individually expensive and on which the mark-up is traditionally high. If he foresees sufficient turnover, therefore, he will be able to outbid the butcher. The latter may settle for a less important and therefore less expensive position in the shopping centre whilst a carpet or furniture retailer who requires quite large premises will tend to seek premises some distance from the prime part of the centre. The trader is likely to choose the best pitch within his financial ability and any analysis of rents agreed will provide information on traders' preferences.

Where a particular unit has been let, it is likely that companies representing a range of retailing activities have competed to obtain a lease of the premises. As a result the rent achieved is strong prima facie evidence of the current market rental value.

Whilst the retailer's maximum rent is governed by the level of profit he can generate he will compete in the market and offer that maximum rent only where the competition is sufficiently strong for him to do so.

Table 11.1 Typical percentage shares of turnover payable to landlord as additional rent, the base rent having been agreed, typically at 80% of the full market rent

Percentage	Trade
1.5	Grocery store
3	Department store
5	Furniture
5.5	Restaurant
7	Electrical goods
7.5	Men's fashions
8	Books, sports goods
9	Ladies fashions
10	Leather goods, jewellery

Source: D. Andersen (adapted)

The percentage on turnover will be affected by the anticipated mark-up of the goods sold and the volume of sales anticipated. Where margins are low and trade competitive, as in grocery stores, the percentage share may be as little as 1.5% of turnover (although the high volume of trade may compensate for the low percentage).

At the other end of the scale, where luxury goods are sold with high margins, the percentage share payable to the landlord may be 10% or more.

11.1.1 Turnover rents

In the USA and to a lesser extent in the UK, the actual trading result is sometimes made the basis of the rent assessment. The most common modern form is an arrangement whereby the tenant agrees to pay a basic rent of 80% of the rack rental value together with a predetermined percentage share of his actual turnover, or turnover above a certain level (together termed a 'turnover rent'); a share that will vary from one trade to another — a range of typical percentages is shown in Table 11.1. The landlord therefore relies on the anticipated profitability of the location to make up the basic rent to at least the level payable under a traditional lease at a rack rent.

In proceeding in this way, the landlord has a direct interest in his tenant's level of trading and accepts an additional and entirely different form of risk and one which is usually shouldered by the tenant, since he is closely locked into actual performance, whether due to the activities of the tenant or the wider economic circumstances. Among other things he has to weigh the attraction of what amounts to a yearly rent review (though not necessarily upward only) against exposure to the vagaries of trading.

The method was developed in the USA in the depression of the 1930s when shops were likely to remain vacant unless the landlord was prepared to accept rent based on the level of trade taking place. The system is no longer grounded in such desperate conditions. The method is not widely used in this country, although one major developer operates turnover rents as a policy. It may be argued that the format is fairer to both parties in the early stages of a shopping development where it is more difficult to determine rental values since there is no rental evidence and only limited trading experience on which to base any estimate. Both landlord and tenant have a vested interest in the success of the venture and it may be that the landlord will respond more quickly and positively to the needs of the tenant where he has a direct pecuniary interest in the outcome.

More recently, British Rail have recognized the monopoly value of its main line termini forecourts and offered site, kiosks and small units on licence (as opposed to lease) and with turnover rents. In this way it shares in the prosperity of the site at which it picks up and deposits many thousands of passengers. By use of licences it is able to control the activities of traders more closely and change or discontinue uses where appropriate.

11.2 OTHER COMMERCIAL USERS

Location is an important consideration for most businesses but is extremely critical in the case of retail units. The valuer has access to far more information when dealing with the value of office blocks and industrial units where the relevant information may be gathered from a larger area and still give valuable help in determining the level of rents.

11.2.1 Quasi-monopolies

There are circumstances where the business premises stand alone or in some other way afford a quasi-monopoly. Any such monopoly is likely to be related to some extent to location, statutory (including planning) restrictions and/or licensing or franchising schemes which have the effect of restricting the number of businesses of the same type operating in the same area.

Types of business to which the description applies include hotels, restaurants and public houses, petrol filling stations, sub-post offices, betting shops, amusement arcades, gaming clubs, car parks, leisure activities and various franchises.

Apart from planning requirements there are no restrictions on siting petrol filling stations adjacent to one another although, since it would not make economic sense, the filling station can be seen as enjoying an element of monopoly in the immediate area. A petrol filling station on a busy urban dual carriageway would be likely to have a larger throughput of petrol and other sales than a filling station of similar design on a B classified road between two small towns. But location is not the only factor; hours of opening, pricing policy, the brand and grades offered and the availability of other facilities such as an automatic car wash unit and valet service will reflect in profitability and therefore in the level of rent which the operator is able to pay or price which a purchaser is prepared to pay.

11.3 THE USE OF TRADING ACCOUNTS

Both rental and capital values tend to be directly influenced by the potential for profit and in these circumstances a valuation having regard to the profits achieved is more likely to produce a realistic valuation than any application of comparison methods.

The process is somewhat crude and not carried out with the degree of precision expected by an accountant. On the other hand where used by a valuer experienced in the particular type of trade or business activity under consideration it can provide a useful guide to value. The method is widely used by specialist valuers and involves adjusting the net profit as indicated in Figure 11.1 to find the 'divisible balance', part of which is allocated to rent which is then capitalized in the usual way where a capital valuation is required. The value of the business is obtained by applying a multiplier to the goodwill, the total value of the business being the value of any interest in property and the value of the goodwill with additional payments for equipment, fixtures and fittings.

Alternatively, the net profit alone may be used to estimate the value of the business 'lock, stock and barrel' (also shown in Figure 11.1).

11.3.1 The approach in more detail

Where possible the valuer uses figures derived from recent accounts which will enable him to measure the sums involved in achieving the annual turnover and the level of expenses incurred before gross and net profit figures are struck.

A well established practice is to look at the accounts for the three most recent trading years. A series of accounts will give a better indication of the general health of the business than could be obtained from a set of accounts for one financial year only.

Given sets of accounts for two businesses carrying on a similar activity and each producing the same average level of net profit over the 3 years, the valuer is more likely to be impressed with accounts which show a steady improvement in turnover and net profit than with one where the net profit remains constant or even falls.

The inability to produce accounts for the most recent year should be viewed with some suspicion and also where the accounts appear to be

Profits method
The two approaches in common use are:
1. Goodwill
   Turnover   £
   less purchases
         expenses of trade
         interest on tenant's capital (fixtures and fittings, cash requirement,
         remuneration)
   Divisible balance (for allocation between rent and profit)   £
      Rent allocation   £
      YP (at property investment yield)
      Capital value   £
   Add Goodwill
        Adjusted net profit
        YP (multiplier usually 1 to 5)   £   £
   Value of freehold and business as going concern   £

2. Lock, stock and barrel
   Estimated profit
   (turnover less costs and expenses as above)   £
   YP (modest, reflecting intangible nature)
   Estimated rental value   £
   YP (at property investment yield)
   Value of freehold and business as going concern   £

Note
Both examples assume that the freehold is being acquired by the purchaser of the
business. When the business is associated with a leasehold interest only, it would be
normal to assess the profit rent and arrive at the value of the unexpired term. There is
often a value associated with the purchase of a leasehold interest, beyond the
capitalized equivalent of the profit rent. The additional value reflects the security
offered to occupying tenants by the provisions of the Landlord and Tenant Act 1954.
   In the case of public houses, the brewer may be prepared to charge a lower rent in
exchange for an undertaking by the licensee to sell only the beers, etc. produced by that
brewer. The brewer thereby ensures a market for his product on which he expects to
make a profit. The lower rent is referred to as a 'tied' rent and the profit rent may be
seen as the difference between the market rent (free of tie) and the tied rent.

Figure 11.1 Valuation approaches by the profits method.

subject to unreasonable delay, which may be not unconnected with the
reason for sale. It may be possible to gain some insight of those accounts
from the owner's accountants (but only of course with the owner's co-
operation). It should be emphasized that there is no substitute for the
audited and certified accounts.

11.3.2 Other enquiries

External checks should be made wherever possible to verify the main
parts of the accounts. The valuer should be vigilant in subjecting the

accounts to critical scrutiny. Where it is known that the profit made on a certain item by an efficient entrepreneur is approximately 30%, then the valuer is entitled to be wary of accounts which appear to suggest an appreciably higher profit. For example, where an activity forms part of a larger group, it may have been the practice to supply all purchases from another member of the group at favourable rates. Needless to say, such a facility cannot be relied on by a purchaser of the venture who will need to budget for full costs.

Quite often an owner will direct his business without charging a salary to the accounts, whereas to find the true profit level the accounts need to be debited with an appropriate amount for the management input.

11.4 POTENTIAL

A prospective purchaser may hold the view that current profits could be improved substantially by better management, improved financial controls, the incorporation of other sales lines or improvements to the premises. To the extent that any potential would also be recognized by other prospective purchasers, the ability to create additional profits will be taken into account by interested parties in assessing the current value. Such aspects should be reflected as an intrinsic part of the value of the business, although the eventual purchaser would expect some benefit (i.e., he would not be prepared to offer the whole of the increase in value attributable to the improvement). In fact, there may be a perfectly good reason why the vendor has not implemented the change (or perhaps did so in the past and found it unsuccessful). The purchaser will be taking the risk of failure associated with any change or new idea.

11.5 SPECIAL FEATURES

An important part of the value of any type of business to which the profits method might be applied is the personality of the owner or operator. It is well known that a business fronted by a celebrity or character is likely to experience an increase in custom. For example, a famous chef will be of immense value to a restaurant. Ex-boxers are often very successful public house landlords. The same may apply to experts; a car tuning station run by an ex-racing driver or a sports centre by an Olympic athlete are likely to attract additional business by association with the well-known name. Such an aspect of value is unique and cannot be transferred. Another well-known person could attract his own following, but it would be wrong to pay a price based either on the reputation of the vendor or on the expectations of his own drawing power.

Conversely, it is sometimes found that a business has been neglected by an owner who is financially restricted, ill, lazy or incompetent. It may be judged that such a business is capable of substantial improvement given new unfettered management. To the extent that it can be judged, some allowance may be made for such a prospect, although again the prospective purchaser would expect to negotiate a price which acknowledged the effort, time and risk associated with the upgrading.

11.6 BUSINESS ACCOUNTS

The use of accounts in arriving at the value of the business and possibly the rental value suggests that the valuer requires an ability to look at a set of business accounts and to have an understanding of their implications.

The method involves much more than selecting the correct figures and doing the appropriate calculations. The interpretation of accounts will therefore be considered at some length.

However large or small, a business needs various pieces of financial information. The business owner must monitor income and expenditure to be sure of remaining solvent and as an indicator of whether he is charging economic prices to enable his business to survive. The Inspector

Table 11.2   Final accounts: Profit and Loss Account

XYZ Ltd — Profit and loss account for the year ended 31 December 19–1

	19–1 £	19–0 £
Turnover	1 227 500	876 250
Cost of sales	618 750	302 500
Gross profit	608 750	573 750
Distribution costs	(179 000)	(236 500)
Administrative expenses	(189 000)	(299 500)
Other operating income	3 250	1 750
Interest receivable and similar income	4 750	2 750
Interest payable and similar charges	(21 250)	(14 250)
Profit on ordinary activities before taxation	227 500	28 000
Tax on profit on ordinary activities	92 500	5 500
Profit on activities after taxation	135 000	22 500
Dividends paid and proposed	17 500	—
Profit for the financial year	117 500	22 500
Retained profit brought forward	30 000	7 500
Retained profit carried forward	147 500	30 000

Table 11.3   Final accounts: balance sheet

XYZ Ltd — Balance sheet as at 31 December 19–1

	19–1 £	19–0 £
Fixed assets		
Intangible assets	38 250	25 250
Tangible assets	503 750	346 500
	542 000	371 750
Current assets		
Stocks	109 250	199 500
Debtors	219 750	197 750
Cash at bank and in hand	50 250	20 000
	379 250	417 250
Creditors: amounts falling due within 1 year	152 500	145 000
Net current assets	226 750	272 250
Total assets less current liabilities	768 750	644 000
Creditors: amounts falling due after more than 1 year	234 250	227 000
	534 500	417 000
Capital and reserves		
Called up share capital	375 000	375 000
Share premium account	5 750	5 750
Revaluation reserve	6 250	6 250
Profit and Loss Account	147 500	30 000
	534 500	417 000

of Taxes and Customs and Excise will expect to find financial records in any business (they are required by law for companies, whether private or public) and it is much easier to deal with questions, tax liabilities and other matters of that nature if the accounts are well-maintained and up to date.

Where outside funding is required for acquisition, the lender will wish to investigate the financial integrity of the business before committing himself; the cost of finance may well be influenced by the way in which the accounts are presented as well as by the picture they give of the business.

Accounts fall into two main groups: those concerned with the performance of the business over a particular year or other accounting period (to which the Profit and Loss Account refers) and those concerned with the capital assets and structure of the business (dealt with in the balance sheet). The groups will be discussed separately with reference to the set of final accounts appearing in Tables 11.2 and 11.3.

11.6.1 The Profit and Loss Account

The Profit and Loss Account is prepared from a Trial Balance, which is a list of all the accounts held for the year in question showing their positive or negative balances. The appropriate items are extracted and used in the final accounts.

It should be appreciated that the final accounts are summaries of the company's operation, which are supplemented by notes to the accounts. The more important and usual items in the Profit and Loss Account or in the notes include:

Cost of sales
Turnover
Net profit before tax
Depreciation*
Machinery and plant hire*
Auditors' fees*
Interest on loans*
Directors' remuneration and pensions*
Non-recurring losses

together with income in the form of:

Income from investments
Rents received
Non-recurring profits and also
Corporation Tax

* Will appear in notes to the accounts as shown in more detail in Figure 11.2.

The Profit and Loss Account shows the net profit after tax following which the profit is appropriated. The Profit and Loss Account in Table 11.2 is a summary of the year's trading activities. It has several items of interest on which further comment will be made.

The second column shows the comparative figure for the particular item for the previous year.

The gross profit is obtained by deducting the direct cost of sales from the turnover. The gross profit is almost 50% of total sales whereas in the previous year it was 65.5% on an appreciably lower turnover. An examination of the various costs involved would perhaps reveal whether this result was exceptional or likely to be continued and also indicate whether further increases in turnover would involve a lower percentage of gross profit.

The net profit before tax is derived by deducting all indirect expenses from the gross profit. (Any interest, rents or other payments received will be added back.) The net profit shows a return of 18.5% on turnover and 37% on gross profit (previous year 3.2 and 4.9%, respectively).

Profit and Loss Account
  Wages and salaries
  Social security costs
  Depreciation
  Auditors' remuneration
  Interest payable on borrowings for periods of 5 years and under
  Interest payable on borrowings for periods exceeding 5 years
Details of directors' emoluments if total exceeds £60 000
Particulars of employees
  (average numbers)
  (geographical split)
Details of higher paid employees (exceeding £30 000 p.a.)
Details of make up of tax figure

Balance sheet
Breakdown off intangible fixed assets, e.g., goodwill, patents, etc.
Breakdown of tangible fixed assets
  Land and buildings

Plant and machinery
Fixtures, fittings, tools and equipment
Breakdown of stock
  Raw materials
  Work in progress
  Finished goods and goods for resale
Breakdown of debtors
  Trade debtors
  Prepayments and accrued income
Breakdown of creditors up to 12 months
  Bank loans and overdrafts
  Trade creditors
  Taxation
  Proposed dividends
  Accruals and deferred income
Breakdown of creditors exceeding 12 months
  As above and
  Debenture loans
Called up share capital
  Ordinary shares
  Preference shares

Figure 11.2 Items to be found in notes to accounts.

An investigation is needed to find the underlying cause of the high distribution costs and administrative expenses in the previous year.

The way in which the profit is allocated between dividends and transfer to reserves will be determined by the needs of the owners and the nature of the operating company. It may indicate the policy of the company towards ploughing profits back into the business.

11.6.2 Stock

Stock acquired for resale but not yet sold is normally recorded at cost price (but stock which has no market because it has been superseded or outlived its shelf life would not be included except at net realizable value or scrap value (if any). Practice will vary dependent upon the type of business and the accounting conventions employed.

Where a substantial amount of work to raw materials takes place to convert into a component or finished goods, it is usual to include the cost of the work in the stock value although it would be misleading to include the goods at their anticipated sale price, thereby anticipating the expected profit on sale.

Debtors are recorded as assets unless bad debts are anticipated. Payment of an outstanding account will reduce the total figure attributed to debtors and increase the bank balance by a similar amount.

The bank balance recorded shows the amount held in the bank for normal trading purposes: the firm would not expect to keep more than is necessary for day-to-day operation because it should be better employed in the business.

11.6.3 The balance sheet

The balance sheet displays all the assets of the business including any profit or loss not distributed from previous accounting periods. Table 11.3 shows a typical balance sheet for a medium-sized firm. The total amount shown is the theoretical value of the business — the amount available should the assets be sold. A business in sound heart would be expected to realize more than is shown for several reasons. First, there is a tendency to write down assets for depreciation (and tax) purposes, regardless of their true value in the business or the market. After some time, many of the more durable assets of a business will have a nil book value although remaining in operational use. Second, purchasers would consider making a payment for goodwill, which may be described as the opportunity to continue trading with the company's current customers and acquiring new customers attracted by the firm's name, trademarks and reputation.

There is no unanimity at the present time concerning the treatment of goodwill in accounts — the policy of some firms is to exclude it or write it off whereas other firms believe it should be shown in acknowledgement of the important part it plays in the success of the business.

Each item shown on the balance sheet will be considered in turn.

Assets
Fixed assets include both tangible and intangible assets and investments: tangible assets include land and buildings, plant and machinery and fixtures, fittings, tools and equipment; intangible assets represent the company's view of the value of its goodwill, patents, licences, trade marks and similar rights and assets. As mentioned earlier, there is no current uniform practice in relation to goodwill but there should be no objection to its inclusion where there is some evidence that a value exists. The valuer should be aware that a figure is included and may need to form a view as to whether it is a reasonable amount.

The net current assets are represented by stock, debtors and cash less the total amount owed by the company to short-term creditors, for example, trade creditors, bank overdrafts and tax owed.

Issued share capital
Capital may be provided initially by the trader, his friends and relatives, by a bank or other lending institution or may be raised by public subscription or it may be classified as share capital (where the subscribers

rely for their return on the profitability of the business as interpreted in dividends fixed by the annual meeting of shareholders of the company). It may be provided through debentures or other loans when the terms will be based on commercial considerations and payment of interest is due regardless of the company's results.

Profit

Unless all the profits are distributed to those having shares in the business, the balance retained will be shown in the balance sheet and contribute towards the reserves of the company.

The balance may be recorded in the profit and loss account and may be added to the general reserves or paid as dividends.

Share capital may be increased or decreased as may be debentures. The example shows that shareholders' funds are the same as in the previous year. Repayment of debentures may take place to meet an obligation to repay by a certain date or the management may make a repayment based on its judgement that it would be cheaper or otherwise expedient to place more reliance on share capital (as pointed out above, there is no fixed return on share capital, the rate of return depending largely on the health of the business and the dividend policy).

Stock acquired for resale but not yet sold is normally recorded at cost price (but stock which has no market because it has outlived its shelf life or has become obsolete would be included, if at all, at its scrap or net realizable value).

Other assets

In addition to the set of accounts which must be audited and certified there must be a Directors' Report containing further statutory information. The financial information given in the accounts and the accompanying statements is the raw material on which the valuer may draw to prepare his assessment of the viability, profitability and exchange value of the business.

11.3.4 Performance indicators

It will be useful to consider a number of accepted measures of performance employed by accountants and others in assessing the financial health of a business.

Both the profit and loss account and the balance sheet may be analysed in a number of ways. Where more than 1 year's accounts are available, it may be possible to deduce other information about the business or it may suggest other enquiries which should be made.

Accountants use a number of ratios and percentages to help them draw conclusions from their clients' accounts, enabling them to refine the quality of advice given to clients. Ratios have varying levels of significance and are used in different circumstances. The main ones likely to be of use to valuers are included in Appendix 1.

Adequate working capital is important. Insufficient capital will inhibit the growth of the business and perhaps even threaten its existence. A rule of thumb is that provided by the current ratio, where it is generally accepted that a ratio of 1.5 to 2:1 is reasonable whilst any higher ratio should be explored since it is possible that too much in the way of assets is lying idle.

In the balance sheet shown in Table 11.3, the current assets (stock, debtors, bank) amount to £379 250 whilst the current liabilities amount to £152 500, giving a ratio of 2.49:1, showing the company to have a strong liquidity position.

12

Application of the profits approach

The profits method is much used by specialist valuers in the hotel, filling station and leisure industries for valuation and sale purposes. It is also employed in appropriate circumstances by the Lands Tribunal in compensation claims and by the Inland Revenue in assessing rateable values. The following examples show several types of property. One feature which is very evident is the way in which the Lands Tribunal tends to apply a very simplified version of the approach. Reasons have not been given but the Lands Tribunal has tended to avoid any approach which it regards as too academic or intricate. Perhaps it is not surprising that the Tribunal is able to report its decision quite simply when it is, in effect, seeking to distil the more likely components of value from sets of figures and evidence provided by two specialist valuers.

Example 12.1

Petrol filling station (as reported in *Payne* v. *Kent County Council* [1986] 280EG.645)

The Lands Tribunal member considered a compensation claim following compulsory acquisition of the petrol filling station, workshops and cafe adjacent to a bypass within a quarter of a mile from junction 3 of the M2. The Tribunal found itself adjudicating between petrol sales estimates of 1 100 000–2 000 000 gallons and capital values from £500 000 to £1 135 000. The Tribunal arrived at a figure of £750 000 made up as follows:

	£
Petrol filling station — estimated petrol sales (gallons p.a.)	1 500 000
at a rental of 3.5p per gallon	3.5
Annual rental	52 500

YP perpetuity at 6%		16.67
	say	875 000
less cost of redevelopment		175 000
Capital value		700 000
add to reflect opportunity to provide		
small catering unit		50 000
Compensation awarded		750 000

Commentary
The major part of the value is attributable to the favourable location of the filling station. The petrol throughout and the estimated rental value of 3.5p per gallon is used to determine an annual rental value which is then capitalized. A lump sum is added to reflect the possibility that a small catering operation would be possible.

Example 12.2

A Lands Tribunal decision on the value of a petrol filling station acquired for a local road scheme (*Mayplace Garage (Bexleyheath) Ltd* v. *Bexley London Borough Council*, Ref 33/1988).

The matter was complicated because two sites were involved, the filling station site owned by the company and an adjoining site owned personally by the managing director of that company.

Evidence was given that the existing station was outmoded in layout and appearance and in disrepair. The Tribunal accepted that the sites could be merged for the purposes of assessing compensation. The Tribunal assessed the capital value in the following way:

	£
Estimated annual sales volume in gallons	750 000
Capital value at 0.75p per gallon	562 500
deduct estimated cost of redevelopment	200 000
Capital value (free of oil company supply agreement)	362 500

Commentary
The anticipated volume of petrol sales was again used but in this case the unit figure used was directly related to capital value. The sales were those anticipated of a newly developed filling station and the estimated cost of new buildings.

The value of the redeveloped site was assessed as being free of tie to any oil company.

Example 12.3

Sub-post office and general store

Modern shop premises at the end of a terrace in a busy local shopping area in a popular suburb some 3 miles from the city centre are used to operate a busy sub-post office and small store. The first floor accommodation is used as storage apart from one room used as an office. The Post Office salary is £30 000 p.a. whilst the store has a turnover of £1500 per week.

The business is for sale as a going concern together with the freehold; the Head Postmaster is prepared to transfer the appointment to a suitable applicant.

Commentary

No doubt the value of the vacant shop can be compared with other transactions in the area relating to similar shop premises. The Post Office appointment is a quasi-monopoly and is likely to be assessed at a multiplier of the net profit arising from that operation. The multiplier varies, typically between three and five but dependent upon the type of area, the opportunity to increase the volume of service and the likelihood that the appointment will continue into the future. There may be an additional value due to the trade carried on in the store portion of the premises but this will depend on whether the existing trade is likely to continue and to what extent it exceeds the trade that a retailer setting up in vacant premises would receive. The existence of the sub-post office section is often extremely helpful to the other trade carried on in the premises especially where it can be seen as complementary, for example, a store selling greetings cards, gifts, confectionery and items of stationery.

Example 12.4

A public house and wine bar occupy a prominent corner site in the retail centre of a prosperous market town. In addition to the sales of alcoholic drinks, catering is carried on in the wine bar section and there are two gaming machines. The example shows an outline approach to the ascertainment of rateable value on which is based the assessment of the national non-business rate. A similar process could be used to arrive at the market rental value.

	£
Draught beer — 600 barrels at £12.00	7 200
Bottled beer — 75 barrels at £15.00	1 125
Wines and spirits — 1500 gallons at £2.60	3 900
Brewer's (wholesale) profit	12 225

plus
Estimated tied rent 600 + 75 + 1500/3
1175 converted barrels at £8.00   9 400

	£	21 625
Brewer's bid — 50% of £21 625	10 812	
Two gaming machines	2 500	13 312
Catering receipts (excluding VAT) at 10%	186 000	18 600
Net annual value	31 912	

Commentary
The sales are reduced to equivalent barrels, which simplifies assessment
of the relative contribution of each type of drink. It is then assumed that
a brewer would be prepared to pay a percentage — most often taken at
50% — for the opportunity to achieve the volume of sales calculated.
Additions are then made for other trade such as catering and accommodation receipts.

The contractor's method

PART SIX

Evolution and principles

13

The valuation of a commercial property interest involves the valuer in attempting to predict its rental value and/or capital value at a particular time. This is an activity being frequently undertaken in the UK and indeed in many parts of the world.

Where the valuation is a precursor to a letting or sale in the open market, the valuer is able to refer to his records of recent transactions concerning similar types of property and use the information as a basis.

The conscientious valuer will therefore undertake analyses of all lettings and sales and store that information for future use. In this way, the body of information which is a basic tool of the valuer is developed and refined. As a result, any valuations made by the valuer based on the database of collected information should represent an informed interpretation of recent market activity and therefore one worthy of serious consideration in refining an opinion of the current market value.

These operations are at their most reliable when there is ample investment market activity in both sales and lettings so that there is sufficient evidence available both to establish a broad pattern and to highlight a maverick transaction. Even with such information, the property market is by no means a perfect market; each property is unique and therefore not comparable in precise terms; the participants tend to be secretive and the information is not always reported fully or correctly.

13.1 LACK OF MARKET EVIDENCE

There are also certain types of buildings or uses where the properties rarely change hands. Even where they do change hands, they may form part of a larger transaction of which the property element is only a small part. Under these conditions it is unlikely that an analysis of the exchange price will be helpful.

Most often the properties alone are offered on the open market only when the building has served its purpose and is no longer to be occupied for its present use or the site is to be redeveloped. Schools, churches, public libraries, chemical plants, sewage disposal installations, airports, chemical and oil plants, operational buildings used by statutory undertakers and public property in general will fall into this category. The market for such buildings is therefore likely to be found among those purchasers seeing an alternative use; for example, the conversion of a redundant railway waiting room to an office or a deconsecrated church to an art gallery.

An analysis of such a transaction gives no information about the value of the property devoted to its original purpose. It simply signals the level of price paid in anticipation of the alternative use, tempered as it will be by the knowledge that considerable expenditure may be necessary to make the premises suitable for the intended new use.

13.2 VALUATIONS FOR OTHER PURPOSES

Not all valuations precede a market transaction: there are many instances of buildings or interests where an opinion of value is required even though no market activity will ensue. The most common example is of valuation for rating but there are many others including valuation for balance sheet purposes, for transfer and valuations to support loan applications.

Despite the absence of the type of information which the valuer normally expects to be available for analysis, there is still a need for an opinion of value. He will therefore resort to some other method of formulating a valuation.

13.2.1 A cost-based approach

In this context, one such approach is the so-called contractor's method or basis, a cost-orientated approach. In brief, the method consists of estimating the replacement costs of the building and adding to them the value of the land. It reflects the age and disabilities of a building by reducing the estimated replacement cost; this adjusted value has been termed 'the effective capital value' but the Lands Tribunal (Mr C.R. Mallett) regarded the term as 'a long established misnomer' and preferred to adopt the phrase 'adjusted replacement cost'.

The value of the land is likely to be affected, usually restricted, by the actual building on it. In addition to the broad categories of buildings and uses suggested above, there may be occasions where the information available is extremely sparse or less reliable than usual for whatever reason. In such cases the valuer may decide to use the contractor's basis, either as the only method or as a check on some other method. The first thing to note is that there is a strong cost element in what is intended to be a value judgement. Moreover, the building to be costed is often old and outdated, when cost and value make particularly bad bedfellows. Not surprisingly the approach has been described as a 'method of last resort' to be used only where other methods are inapplicable or impractical. But there are times, such as in the circumstances described above, when there is little or no alternative to the use of this method. The method is well established, having played an important part in rating valuations of certain types of properties for many years.

13.2.2 Company assets

More recently the valuation of company assets has assumed a new importance. In recognition of general concern at the lack of uniformity in the basis of the valuation of property assets by companies, the RICS set up the Assets Valuation Standards Committee (AVSC) which issued its first guidance note in 1976. The notes have been expanded and updated regularly since that time and now provide a comprehensive statement of the standards required. The scope of the valuations includes those for inclusion in Company Accounts, Directors' Reports and other financial statements, for incorporation in prospectuses and circulars and for valuations under the City Code on takeovers and mergers as well as for certain insurance company purposes.

Unlike valuations for rating, valuations for asset purposes have not been litigated and the only exposure to public scrutiny has occurred where valuers for opposing sides in a takeover battle appear to differ by a larger amount than seems credible.

One of the bases of valuation employed for assets is termed 'the depreciated replacement cost basis' described in background paper no. BP3 issued by the AVSC extracts from which are included in Appendix 2. It will be seen that the basic approach to building up a valuation is very similar to the contractor's test about to be described. Asset valuations will be referred to again when the basic test has been described.

13.3 SCOPE FOR DIFFERENCES

The arbitrary nature of the basis, the use of cost as a surrogate for value and the absence of market evidence all suggest that there is considerable scope within the method for differences of opinion.

The levels of rating assessment have been the subject of scrutiny by the

Lands Tribunal whilst the superior courts have often offered valuable guidance on the proper application of the basis.

Consequently, there is more detailed judicial advice and direction than is available for any other method of valuation. This extends not only to the compilation of the costs of building and appropriate allowances for shortcomings but to a consideration of the way in which the capital value produced is converted to an annual value.

13.4 JUDICIAL GUIDANCE

The Lands Tribunal and the Courts have expressed considerable reservations about the basis which gives full rein to a valuer's judgement without offering any market evidence to support the opinion of value against which the results may be judged. But in the Eton College case (see below) the Lands Tribunal held that the contractor's test was the best substantive method in valuing a large public school, provided that the valuer using the method was sufficiently experienced since it was the only evidence available.

That the basis has the respectability of long use is borne out by a remark made by Cave as long ago as 1886:

'interest on cost is a rough test undoubtedly. It is a test in some cases but not a test in others. If the place is occupied by a tenant, it is not a good test at all, because the rent which he actually pays is a far better one . . . But if the place is occupied by the owner himself, then it is in some sense a test, a rough test no doubt, and only prima facie evidence, but still some evidence, to show what the value of the occupation is . . . if he could get the place cheaper, at a less rent than the interest on the cost comes to, it is to be assumed he would not go to the expense of building, he would prefer to take the cheaper course and pay the rent.' (R v. *School Board for London.*)

The Lands Tribunal has criticized the basis for its 'artificiality of approach' and has described it as 'a poor best' (which is no more than is claimed for it; a valuer will hesitate to use the basis — and indeed will have no reason to use it, at least not as a primary method — where he can find some market evidence to enable him to approach the valuation in a different way.)

But the Tribunal has also given some qualified support to the basis, suggesting in the Eton College case that 'the time has come for a good word to be said for this method of valuation . . .'. They went on

'. . . we are satisfied that the contractor's basis provides a valuation instrument at least as precise as any other approach. For this kind of case it is almost certainly the best substantive method that has been devised so far.' (*Eton College* v. *Lane and another*)

which is to go beyond what most practitioners would claim for the basis which is commonly acknowledged to set a ceiling rather than produce a value.

The Lands Tribunal had previously accepted that the contractor's basis 'is sometimes the only practical approach' whilst an elegant and authoritative description (described by Lord Denning MR as the classic explanation) was given by Sir Jocelyn Simon QC when he appeared as Solicitor General in *Dawkins* v. *Royal Leamington Spa Corporation*.

'As I understand it, the argument is that the hypothetical tenant has an alternative to leasing the hereditament and paying rent for it; he can build a precisely similar building himself. He could borrow the money, on which he would have to pay interest; or use his own capital on which he would have to forego interest to put up a similar building for his owner occupation rather than rent it, and he will do that rather than pay what he would regard as an excessive rent — that is, a rent which is greater than the interest he foregoes by using his own capital to build the building himself. The argument is that he will therefore be unwilling to pay more as an annual rent for a hereditament than it would cost him in the way of annual interest on the capital sum necessary to build a similar hereditament. On the other hand, if the annual rent demanded is fixed marginally below what it would cost him in the way of annual interest on the capital sum necessary to build a similar hereditament, it will be in his interest to rent the hereditament rather than build it.'

Lord Denning MR qualified that statement when he said:

'The annual rent must not be fixed so as to be only "marginally below" the interest charge. It must be fixed much below it, and for this reason: by paying the interest charge on capital cost, he gets not only the use of the building for its life, but he gets the title to it, together with any appreciation in value due to inflation: whereas, by paying the annual rent, he only gets the use of the building from year to year — without any title to it whatsoever — and without any benefit from inflation.' (*Cardiff City Council* v. *Williams*)

In *Gilmore (VO)* v. *Baker-Carr* the Lands Tribunal enunciated a set of rules presented as five stages and which have found general favour.

1. First stage — Estimate the cost of construction of the building.
2. Second stage — Distinguish between value and cost and make deductions from the cost of construction to allow for age, obsolescence and any other factors necessary to arrive at the effective capital value.

3. Third stage — Establish the cost of the land.
4. Fourth stage — Apply the market rates to decapitalize the capital value.
5. Fifth stage — Take into account any items not already considered.

The Tribunal member (Mr C.R. Mallett) added a sixth stage in reporting his decision in the Imperial College case. He said

'I hesitate to add another stage to the saga of the Contractor's Test but I think it is logical and necessary first to determine the annual equivalent of the likely capital cost to the hypothetical tenant on the assumption that he sought to provide his own premises, then to consider if the figure is likely to be pushed up or down in the negotiations between a hypothetical landlord and a hypothetical tenant having regard to the relative bargaining strengths of the parties.' (*Imperial College of Science and Technology* v. *Ebdon*).

He acknowledged that some of these factors had been listed by Lord Denning MR in the Cardiff City Council case where, in typical fashion, he considered the matter in helpful detail. He enumerated some of the many relevant factors.

1. The hypothetical landlord would be anxious not to leave the premises empty, bringing in no return on his capital. He would be content to accept much less than the 6% which the landlord of commercial premises would expect.
2. The hypothetical landlord would be providing the premises for the purposes of the public good, viz., for educational purposes, and would be disposed to accept such rent as the hypothetical tenant could afford to pay without crippling him or his activities, especially if he was a non-profit making body.
3. On the one hand, the hypothetical tenant might be very anxious to get possession of the premises. If he was under a statutory compulsion to establish such a college, he would feel that he must have them — not perhaps at any cost, because if the rent was too high he would look elsewhere — but at such reasonable rent as he could persuade the hypothetical landlord to accept. If he felt a moral compulsion to establish such a college — out of his public duty — he would probably pay as much as if he were under a statutory duty.
4. On the other hand, the hypothetical tenant might wish to have the premises — as being desirable — but not as being compulsory, not as a matter of urgency. He would be inclined, therefore, to hold back and pay less than a hypothetical tenant who felt a pressing need to get them.
5. The hypothetical tenant would, of course, have to consider the means at his disposal. If he was a government establishment, such as a local

authority, supported entirely by Government grants and the rates, he would count the cost, but not perhaps so anxiously as a non-government establishment under a pressing necessity to make both ends meet. (*Cardiff City Council* v. *Williams*).

13.5 DECAPITALIZATION RATES

As noted earlier, rating valuations are concerned with annual value. Much has been said on the process of conversion to capital value. The applications are looked at in detail in Chapter 14 but it is worth pointing out that the decapitalization rates do not follow those used in commercial transactions.

The application of the rules represents a theoretical basis for a reasonable attempt at valuation in those limited circumstances where it will be employed. Given that there are circumstances when the method will be employed, the important need seems to be a proper understanding of the principles, coupled with sound information.

13.6 COSTINGS

As shown earlier, estimated cost forms an important part of the test. The estimate may be approached in one of two ways.

In the first approach, an estimate of the cost of replicating the building broadly in its existing form is prepared from which are deducted allowances for age, obsolescence, excessive ornamental treatment and the effects of bad layout, design and any other aspect where the building is deficient in modern functional terms. This is referred to as the 'adjusted replacement cost'.

The second approach is to estimate the cost of what has been described as the 'simple substituted building' which retains the essential functional nature of the original without its ornate and expensive style and structure. Again there would be deductions to reflect the age and obsolescence of the actual building.

Both responses reflect the likely attitude of a commercial occupier faced with the need for certain accommodation. Whether leasing or purchasing the building, the occupier would be likely to consider the cost and convenience of a modern functional building over an older obsolescent one; given a choice unrestrained by what actually exists, he would tend to occupy the turn-of-the-century building only if the total costs of occupation (including such important items as depreciation, maintenance and heating) were roughly the same.

The various components in providing an estimate relate to the costs of building, the effect of age and obsolescence and the value of the land. These matters will be considered in turn.

13.6.1 Building cost

The first requirement is to estimate the cost of constructing the existing building at current prices. The task is complicated by the fact that construction methods and building materials have changed over the years. It should be remembered that what is required is an approximate cost rather than a precise cost for it is to be used in a method that is regarded as a not very reliable when used on its own. But where the valuer is concerned that he is outside the boundaries of his own knowledge and expertise, it may be worthwhile to make use of the services of a quantity surveyor.

In describing the different methods of estimating available it is again emphasized that the requirement is to determine an approximate cost. There are five methods commonly in use.

Unit cost

The unit cost, depending on the type of building being priced, will reflect the overall quality of construction and include allowances to reflect special finishes, lifts, heating, air-conditioning plant, etc. Various units may be assessed by applying a unit rate. Schools, cinemas, hotels and hospitals, for example, may be assessed in this way; schools by cost per place, cinemas by cost per seat and hotels and hospitals by cost per bed. The figures will of course vary according to the quality of the building and of its finishes.

Cubic content

By reference to a unit figure applied to the total cubic content of the building. The calculation is based on external measurements, the external perimeter measurements of the walls being used in conjunction with the height. Foundation costs and the difference in cost between a flat roof and various designs of pitched roofs are reflected by arbitrary additions to the measured height of the building.

The precise procedure for measurement forms a standard promoted jointly by The Royal Institution of Chartered Surveyors and The Incorporated Society of Valuers and Auctioneers. The relevant section is reproduced in Appendix 3. The cubic content approach is now used less in preference to more sensitive information available in respect of units of floor area. The approach would still be useful in preliminary costings or in the case of a building having considerable height without intermediate floors, as for example a church.

Superficial area

By far the most used method of approximate estimating replacement costs is the superficial area method. The gross internal floor area is calculated (total space within inside faces of external walls) and a unit price applied with lump sum additions for car parking, stores and other structures outside the main building. That the method requires a degree of experience and skill is borne out by the wide range of costs encountered on this basis; a recent example indicates a cost of £900 per m^2 in respect of a church and a cost of £1400 per m^2 in respect of a cinema (including fitting out with equipment and providing air conditioning plant).

The effect of plan shape on price is not reflected directly in this approach and the valuer should be aware that departure from a simple square or rectangular plan is likely to have an upward effect on prices. Costs tend to be affected to some extent by regional variations and various cost references are available. The Building Cost Information Service (BCIS) is a subscription service provided by the Royal Institution of Chartered Surveyors which gives quite detailed cost breakdowns of a limited number of recent building contracts (Table A.1 is a typical example). One of the best known books of prices is that published by Spon which includes a summary of the unit price range for a variety of buildings and materials (Figure 13.1 shows a specimen page). Crosher and James, Chartered Quantity Surveyors, publish at intervals a card which provides average price ranges for various types of residential, commercial and civic administrative buildings. The card (Figure 13.2) also details regional variations throughout England, Wales and Scotland.

Elemental costs

In assessing cost by reference to elements the material and labour costs for each main component of the building are identified and totalled to arrive at an overall cost. This approach has gained in popularity since the adoption of cost planning by many quantity surveying practices. It requires a more detailed knowledge of construction than either of the two former methods and will not necessarily prove any more accurate. Published information on elemental costs is available from BCIS. Appendix 4 gives some typical information.

Approximate quantities

Quantities may be taken off, priced and totalled. In this way the cost of the main materials finishes, fixtures and equipment together with the labour operations involved are accumulated and converted to a cost per square metre to which are added lump sums for external and other work not included in the initial costings. This method is the one in common use by quantity surveyors but is likely to be beyond the experience or ability of the valuation surveyor.

Administrative, commercial protective service facilities (CI/SfB 3) - cont'd	Square metre excluding VAT £
Large trading floors in medium rise offices	1800 to 2100
Two storey ancillary office accommodation to warehouses/factories	565 to 630
Fitting out offices	
basic fitting out including carpets, decorations, partitions and services	180 to 225
good quality fitting out including carpets, decorations, partitions, ceilings, furniture and services	360 to 435
high quality fitting out including raised floors and carpets, decorations, partitions, ceilings, furniture, air conditioning and electrical services	630 to 785
Office refurbishment	
basic refurbishment	300 to 450
good quality, including air conditioning	600 to 800
high quality, including air conditioning	1100 to 1450
Banks	
local	925 to 1100
city centre/head office	1400 to 2000
Building Society Branch Offices	800 to 1050
refurbishments	450 to 750
Shop shells	
small	390 to 495
large including department stores and supermarkets	360 to 405
Fitting out shell for small shop (including shop fittings)	
simple store	400 to 465
fashion store	750 to 925
Fitting out shell for department store or supermarket	
excluding shop fittings	450 to 585
including shop fittings	675 to 925
Retail Warehouses	
shell	250 to 370
fitting out	180 to 210
Shopping centres	
Malls including fitting out	
comfort cooled	1350 to 1500
air-conditioned	1500 to 1700
Retail area shells, capped off services	400 to 500
Landlord's back-up areas, management offices, plant rooms non air conditioned	560 to 630
*Ambulance stations	520 to 750
Ambulance controls centre	750 to 1100
Fire stations	750 to 1050
Police stations	775 to 1150
Prisons	1050 to 1250

Health and welfare facilities (CI/SfB 4)

*District general hospitals	825 to 1100
Refurbishment	400 to 800
Hospice	850 to 1050
Private hospitals	900 to 1250
Hospital laboratories	1050 to 1500

Figure 13.1 A selection of unit building costs from Spon's *Architects' and Builders' price book 1991.*

DATACARD

BUILDING COSTS CARD
JULY 1989   ©

CROSHER & JAMES
CHARTERED QUANTITY SURVEYORS
INTERNATIONAL COST CONSULTANTS

GUIDE PRICES PER SQUARE METRE OF GROSS INTERNAL FLOOR AREA FOR NEW BUILDINGS AS AT JULY 1989

The following are average price ranges for constructing buildings in South East England exclusive of professional fees, fitting out costs and V.A.T

OFFICES	£/m2	RESIDENTIAL	£/m2
Low rise basic	575-850	Local authority flats	425-600
Low rise air-conditioned	800-1100	Local authority houses	375-500
Medium rise basic	700-950	Private flats	450-575
Medium rise air-conditioned	950-1250	Speculative houses	500-750
		Luxury flats	675-950
High rise basic	950-1350	Sheltered accommodation	425-550
High rise air-conditioned	1275-1775	As above with warden	475-625
Prestige	1275-2300		

RETAIL		INDUSTRIAL	
		Light industrial basic	225-375
Shop shells	300-500	Light industrial/offices	325-475
Shells in malls	400-500	Nursery units	350-450
Air conditioned malls	1200-1700	Workshops	300-450
Shop fit-out	400-800	Unheated warehouses	225-325
Warehouse shells	250-375	High-tech/controlled environment	600-1050
Public houses	625-800		

CIVIC ADMINISTRATIVE		OTHER BUILDINGS	
Council offices	725-875	Three star hotels	725-950
Magistrates courts	800-975	City centre luxury hotels	1250-1550
Police stations	700-950	Multi storey car parks	200-300
Covered swimming pools	950-1200	Laboratories	975-1400
Leisure centres with pool	650-975	Computer buildings	1000-1500

REFURBISHMENT OF EXISTING BUILDINGS

Prices will be dependent upon the extent of works required but can be as high as 80% of the appropriate prices given above.

PROFESSIONAL FEES

For buildings requiring only an architect and quantity surveyor, such as housing and basic industrial developments allow 7.5% - 9%. For more complex buildings requiring a structural engineer and mechanical and electrical services consultants allow 12% - 15% and for more sophisticated developments allow 12.5% - 17% inclusive of V.A.T.

CAPITAL ALLOWANCES

Many buildings contain elements of works which can be eligible for capital allowance tax relief. Crosher & James are experts in this specialised field.

Figure 13.2 Datacard published by Crosher and James.

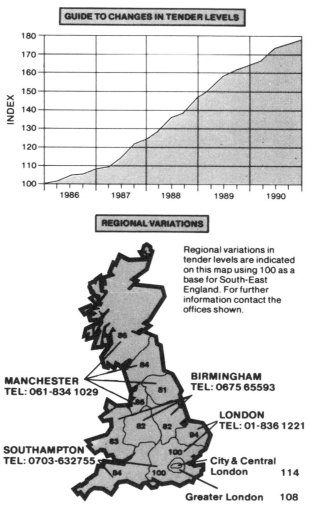

GUIDE TO CHANGES IN TENDER LEVELS

REGIONAL VARIATIONS

Regional variations in tender levels are indicated on this map using 100 as a base for South-East England. For further information contact the offices shown.

MANCHESTER
TEL: 061-834 1029

BIRMINGHAM
TEL: 0675 65593

LONDON
TEL: 01-836 1221

SOUTHAMPTON
TEL: 0703-632755

City & Central London   114

Greater London   108

This Datacard is, by its very nature, simplistic and may not be appropriate for projects with particularly difficult problems in terms of precise location, planning, size, nature of site and time constraints. Professional advice should always be sought at the outset of any building or construction project.

CROSHER & JAMES – Established 1935
1/5 Exchange Court, Strand, London, WC2R 0PQ
Telephone: 01-836 1221. Fax: 01-831 1233.
Offices and Associated Practices in London, Birmingham, Manchester, Southampton, Australia, Singapore, Africa

Figure 13.2 continued

It will be seen that each method demands a knowledge of construction and building costs in addition to a knowledge of the practice of measurement of buildings.

Variation in the quality of construction materials and the standard of workmanship has always existed. Design, detailing and maintenance also play a part in the performance of a building over its life. For these reasons, it is not possible to be dogmatic about the physical life of a building. A well-constructed building built of good materials and adequately maintained may reach the age of 100 years and still have many years of serviceable life. Most of the listed buildings in this country are 150 or more years old; many historic buildings are still in sound condition after some hundreds of years. On the other hand, any building has a finite life and the present age is therefore of some consequence.

The allowance for age in this method is usually combined with that for obsolescence.

13.6.2 Depreciation and obsolescence

The profession has been much concerned about the way in which deterioration in the value of an investment should or could be reflected in any valuation. The suggestions range from the deduction of deferred amounts and reversions to cleared site value to adjustments of the capitalization rate and allowance for regular refurbishment to maintain the quality of the investment. It will be seen from Chapter 13 that the contractor's basis adjusts for these matters through the rate selected which is then applied to a capital value which has already taken into account the fact that the building has been in existence for some time. The subject of depreciation has been discussed in more detail in Chapter 8.

13.6.3 Land

The value of land is usually assessed by use of the comparative or residual methods, depending on the circumstances. The existence of planning permission has an important effect on value but there is a further complication when applying the contractor's test. Any value attributable to the land as a site for development is constrained and needs to be tempered by the fact that there exists a building on the site which may not represent the optimum use or value for that site. The practice has therefore been developed of deducting from the identified site value an allowance to reflect the existence of the particular building on it. Some valuers make the same percentage deduction as that used for the building but that should not be an inviolable rule. The discipline of considering separately the effect on the value of the land of the age and obsolescence of the building is one that should be pursued as likely to be beneficial to the

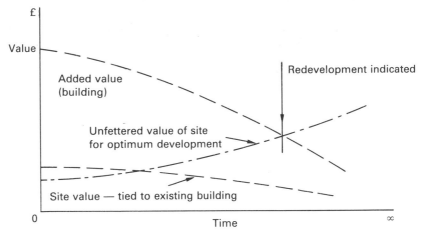

Figure 13.3 Redevelopment decision.

final determination. In certain cases the effect of current planning requirements and/or restrictions may result in the value of the land with a replacement building being less than its value with the existing building.

The interrelationship of the values of the land and buildings will repay some further consideration. One of the benefits claimed for property investment is that rental value is likely to increase by at least the rate of inflation with the expectation that the actual increase will exceed the level of inflation.

Given stable unchanging capitalization rates, the capital value will increase in line with rental value. The total return may therefore be measured in terms of rental income plus capital appreciation. The capital value is an asset comprising land and buildings. It is therefore reasonable to expect that as a building deteriorates through age and obsolescence, the value of the land will deteriorate also where its use is restricted to accommodating the building presently erected on it. The basis of redevelopment is the combination of a deteriorating building and a site no longer fully utilized by that building and which would be enhanced by redevelopment. In approaching the value of the site for inclusion in the contractor's test, the potential of the site must be excluded, since it is anticipated that the existing building will remain. The position may be illustrated graphically (Figure 13.3).

Application of the contractor's method

It is now possible to consider the application of these principles.

14.1 ANNUAL RENTAL VALUE

As has been seen the test throws up a capital 'value' result. In many cases, the valuer will be required to furnish a report as to the capital value but where the valuer is required to give an estimate of annual rental value (as in valuations for rating purposes) further steps need to be taken to 'annualize' the capital value. This is normally done by calculating a return on capital based on the yield expected of the particular investment. Before discussing the actual approach, it is worth looking at the effect of using normal yield principles in assessing value based on the following example.

Example 14.1

Operational premises have recently been completed by a statutory undertaker for its own use. The cost of the land was £500 000 and of the building £3 000 000. The development is of the same size and fulfills the same purpose as an adjoining building erected 60 years ago. The new building has been provided to cope with additional demand; the original building will continue in full operation.

Using the contractor's test, an experienced valuer has arrived at a valuation figure of £2 100 000 for the existing building (applying a 40% discount for age, unnecessary architectural treatment and obsolescence). The statutory undertaker owns all its own buildings and where necessary obtains finance through the Public Works Loans Board. Such loans are normally secured on all the assets and not related to any particular building or other unit. Were an investor to be involved, he would expect a return related to his expectation of future receipts and liabilities. An important feature in fixing the required yield would be the age and obsolescence of the property. As a result of these considerations, he might be prepared to accept a yield of 8% on the modern building whilst requiring 12% on the older one. In that case the annual or rental equivalents would be

	Capital cost/value incl. land (£)	Annual equivalent
New building	3 500 000	280 000
Existing building	2 100 000	252 000

It will be seen that there is very little difference in the resultant rental terms between the ageing building and its modern equivalent. Whilst in capital terms the older building is assessed at 60% of its modern counterpart, in annual terms the older building is shown to produce 90% of the income available from the new building. This is of course attributable to the compensation provided by the yield differential. Even where the return required on the older building is reduced to 10%, the rent would still amount to 75% of that calculated for the new building, against a capital value of only 60% of the new building. The distortion using perceived market rates defeats the efforts made to differentiate between the two buildings by depreciating capital costs. Using the same rate of 8% in each case preserves the differential established by the capital assessment. But this is still a market rate and as has been seen, the method is used because there is no market to observe and therefore no indication of the investment return required by the market, giving no indication of rental value for assessment purposes. The only reason for using the method is that there is no market; there is therefore no evidence of the attitude of investors to this type of funding. The reasons for the choice of particular rates and the reasons for the difference between the rates for various types of building assessed by reference to the contractor's basis are obscure although they are related much more to long-term rates of interest than to short-term trends. The level established for various types and uses of property are shown in Table 14.1. However, this does not explain the discrepancy between commercial interest rates and rates used in the process. It is fair to say that the courts have not been entirely consistent and the view expressed on more than one occasion that the rate to be used should be the commercial rate at which money can be borrowed has been honoured more in the breach than in the observance. Indeed the suggestion is fundamentally unsound because there is no one interest rate even for a hypothetical borrower.

It is also helpful to consider the guidance provided by some of the cases considered by the courts and the Lands Tribunal although some of the

Table 14.1 Levels of decapitalization rates

Type of building	Rate (%)
Public schools University colleges and schools	3–3.5
Local authority undertakings	4.5
Local authority and voluntary schools	5
Industrial activities (profit making) Commercial sports facilities	6
Natural gas operations (Scotland)	7
Oil facilities, pipelines, etc. (Scotland)	7.25

arguments put forward suggest that there are occasions when the rate of interest having been selected is then subjected to an element of post-rationalization.

14.2 DECAPITALIZATION RATES

The comparatively low rates employed have been beneficial to ratepayers and an area for review by the courts and the Lands Tribunal (notably the latter) from time to time.

The following account is a distillation of the evidence given and the conclusions reached in a number of cases where the decapitalization rate was a particular issue.

It is accepted that the starting point is the borrowing rate in the commercial market (itself starting at about 2% above minimum lending rate for those borrowers regarded as 'undoubted'). This is then reduced slightly to find a figure said to be appropriate to all normal cases where ordinary commercial considerations apply and with further reductions where the property was suited to some use by a local authority or where the occupier was a charity.

In the Shrewsbury case which concerned a public school 'not run for profit' it was asserted that

'no tenant would be prepared to offer, nor would any landlord . . . expect . . . a rent based on a commercial percentage of the value of the school. The precise figure which would finally be reached is a matter of conjecture'.

The Lands Tribunal has said

'The rate of interest to be applied, as the criterion of rental value, is not what a contractor would ask but what he would get. It is the rate that would be arrived at after negotiation in the market.'

Then, quoting an earlier decision

'The whole of the circumstances and conditions under which the owner has become the occupier must be taken into consideration, and no higher rent must be fixed as the basis of assessment than that which it is believed the owner would really be willing to pay for the occupation of the premises.'

But as Lord Denning MR observed in the same case

'. . . whether it be 3.5 or 4.5% is all guesswork anyway — you are working in a void.'

In a recent case, evidence was produced to show that in over 100 agreed assessments of universities in England and Wales outside the old LCC areas, assessment was on the basis of a decapitalization rate of 3.5% (all in respect of a valuation date of 1 April 1973).

This case produced a most detailed review of interest rates and their use in valuation cases but of course the information relates to interest rates current at that time. However, a table included as part of the decision (shown in Table 14.2) suggests the appropriate process. Thus, each party starts with the actual minimum lending rate or the average over a selected period and deducts the inflation rate (actual or average for the same period) to find the 'real' rate of interest. A borrower's premium is then added to reflect the likely level above MLR which a borrower would be charged — a 'prime' borrower would pay the lowest premium.

Long-term requirements relate specifically to the rating aspect; it is argued that there is a considerable difference between the long-term requirement of the only possible tenant and the restraints imposed by the term of the hypothetical tenancy. Other factors are then taken into account to find the decapitalization rate. The calculation is finally adjusted to take account of repairs (and depreciation in one case) singularly applicable to rating where net figures need to be converted to gross value.

This detailed consideration tends to confirm that the levels previously adopted are likely to continue in the present era of higher interest rates, since current monetary policy tends to confirm that periods of high interest rates will also be periods of high inflation and that the real rate of interest will vary less than at first sight appears likely.

Regulations made under the Non-Domestic Rating (Miscellaneous Provisions) (No. 2) Regulations 1989 (S.I. 1989/2303) remove the problem of determining the decapitalization rate by prescribing rates for any valuation where the rateable value is being ascertained by reference to the

Table 14.2 Comparison of minimum lending rate and inflation rate

Year	Average minimum lending rate	Inflation rate	Real interest rate	Cumulated real interest rate
1951	2.1	12.1	−10.0	0.7
1952	3.7	6.2	−2.5	1.2
1953	3.9	1.1	2.8	1.4
1954	3.2	4.0	−0.8	1.3
1955	4.3	5.8	−1.5	1.5
1956	5.4	3.0	2.4	1.6
1957	5.6	4.6	1.0	1.6
1958	5.4	1.8	3.6	1.6
1959	4.0	0.0	4.0	1.5
1960	5.3	1.8	3.5	1.3
1961	5.7	4.4	1.3	1.1
1962	4.9	2.6	2.3	1.1
1963	4.0	1.9	2.1	1.0
1964	5.0	4.8	0.2	0.9
1965	6.4	4.5	1.9	0.9
1966	6.5	3.7	2.8	0.8
1967	6.2	2.5	3.7	0.5
1968	7.5	5.9	1.6	−0.2
1969	7.8	4.7	3.1	−0.7
1970	7.2	7.9	−0.7	−1.9
1971	5.9	9.0	3.1	−2.5
1972	5.8	7.7	−1.9	−1.9
Average 1953–72	5.5	4.08	1.42	
Average 1956–69	5.7	3.3	2.4	

notional cost of construction, thereby eliminating any uncertainty or need to negotiate this factor in the determination of the assessment. The appropriate rate is fixed at 4% in the case of an educational hereditament or hospital and in any other case, 6%.

Despite having been described as a method of last resort, the test is in use in the circumstances described and provides some form of ground rules for the valuation of certain types of property. The main use for the test is in the unreal world of rating assessment. It may also be used in some cases of compulsory purchase and in the valuation of certain company assets.

It is not surprising, given the nature and consequences of rating, that many arguments based on the application of the contractor's test have been referred to Local Valuation Courts and the Lands Tribunal for adjudication. It is helpful that the Lands Tribunal decisions are published and that the valuation issues are dealt with at length. It is important that this is so since the Tribunal is the highest authority for this purpose and its decisions cannot be appealed except on a point of law. In addition to a legally qualified President, a number of eminent and experienced surveyors have served and their views are worthy of careful study.

Much of the discussion specifically related to rating matters in the context of the contractor's basis applies also to the assessment of the depreciated replacement cost required for certain asset valuations.

The following examples show valuations for rating purposes culled from decided cases and also asset valuations based on rules contained in the Guidance Notes.

Example 14.2

Independent day school (*Westminster City Council* v. *American School in London and Goodwin*)

Both parties agreed that the school (established by the American community in London after the Second World War to offer a system of education geared to that pertaining in the United States) was to be valued on the contractor's basis.

The effective capital value was agreed at £2 350 000 'to reflect any age or obsolescence allowance that would fall to be considered in the valuation on this basis but not any end allowance'.

The only matters at issue were therefore the appropriate rate per cent to be applied to the capital value to produce an annual value and the amount of any end allowance. For reasons reported elsewhere, the Tribunal decided to assume an interest rate of 5%. In regard to the end allowance for disabilities such as cramped site, high density with limited parking, shortage of playing fields, swimming pools and other recreational facilities, seven different floor levels on the ground floor and stairs and ramps due to the presence of both a sewer and a railway beneath the site the Tribunal deducted 6%. In its view, some of the disabilities were already taken into account in arriving at the effective capital value. The valuation decision of the Tribunal was therefore

			£
Agreed effective capital valuation			2 350 000
converted to annual terms at 5%		117 500	
less for disabilities	6%	7 050	
		110 450	
say gross value	110 000		
Rateable value	91 638		

Commentary

Although the case was a celebrated one and the report detailed, the valuation calculation was of the simplest form. The capital value adduced had, in the view of the Tribunal, already reflected some of the disabilities. Further disabilities were allowed for in an end allowance of 6% of rental value, equivalent to a deduction of £141 000 from the capital value. In other words, the same result would have been achieved by taking 5% of a reduced capital value of £2 209 000 without any end allowance.

Example 14.3

A members' cricket club (*Marylebone Cricket Club* v. *Morley (VO)*)

The Lands Tribunal considered a dispute arising as to the proper valuation for rating of the world famous Lord's cricket ground.

The ratepayer's valuer initially approached the valuation on the profits basis but detailed figures showed that a loss was made; he contended for an assessment to rateable value of £3500 (which by coincidence was the approximate rateable value from 1935 to 1956).

The valuation officer presented a valuation on the contractor's basis:

Valuation 1

					£
Land	playing area	5.25 acres at £4000	£21 000	at 4%	840
	nursery field	5.75 acres at £2000	£11 500	at 4%	460
	car parks	2.5 acres at £ 275	£ 687	at 4%	34
Buildings					1334
			£124 247	at 5%	6212
					7546

say £7500

Valuation 2

Capital value for purposes for which land is used		
16.75 acres at £6000 per acre	100 500	
Annual equivalent at 4%		4 020
Building and site works; cost after making allowance for excess space, height, etc.	409 380	
Annual equivalent at 5%		20 469
Total annual equivalent of land and buildings		24 489
Deduct for disabilities — one third		8 163
Adjusted annual equivalent		16 326

Rent which might be expected to
  be agreed after negotiations
  between landlord and tenant   15 000

This valuation was compared with an earlier valuation produced by a contract valuer in the Valuation Office (remarkable, said the Tribunal, because although they are 'of the same property, on the same date, for the same purposes and on the same side' they were so different in amount). For our purposes it illustrates different approaches to valuation using the same basis.

The Tribunal (Mr John Watson) fixed the rateable value of the hereditament at £9000 but in doing so, rejected the use of the contractor's basis on the following grounds

'. . . we accept [the] proposition that where a hereditament has only one possible tenant its rental value is restricted by the ability of that tenant to pay. That is where the contractor's basis breaks down . . . where the tenant is pleading inability to pay in rent anything approaching what the contractor's method would appear to indicate, the proposition postulates an inquiry into how he is conducting his affairs within the economic limits his policy has imposed. Without abrogating that policy, has he used every endeavour to maintain and increase his income on the one hand and curtail his expenditure on the other? For the purpose of such an inquiry it is well established in law that events the tenant could reasonably have foreseen at the date of the valuation, and the effect they are likely to have had upon his mind, are admissible in evidence.'

Commentary

The two valuations demonstrate different ways of approaching the same problem: valuation 1 assigned differing rental values to three separate areas of land, reflecting their various uses. It then took a 5% return on an estimated cost of the buildings; valuation 2 took an overall value of the land to which it added an adjusted cost for the buildings. Both figures were decapitalized and from their total was deducted a major allowance for disabilities.

Example 14.4

Regional airport (Coppin (VO) v. East Midlands Airport Joint Committee)

The East Midlands is conveniently near the M1.

At the time of the appeal in 1970 it had new and purpose-built terminal buildings and a single main concrete runway. The expenditure on the latter had been greater than was necessary for handling the volume of air traffic at the time and was undertaken with the future in mind. The airport was run by a joint committee of local authorities in the region and they accepted that there would be an annual deficit on the cost of working the airport.

For this reason, the valuation was not made using the profits basis 'there was no profit motive and no profit and therefore the profits basis was clearly inappropriate'.

Both valuers pursued the contractor's basis but produced widely differing results. (Valuation officer £37500, valuer for respondent ratepayers £11400.) The main differences were in the decapitalization rate (5% as opposed to 3.25%) and a deduction for disabilities (15 and 40%).

The decision of the Tribunal member (Mr J.H. Emlyn Jones) reconciled the views of the two valuers (see below). He also accepted that some recognition should be given to what the Valuation Officer first termed 'under-user' but later amended to 'new venture allowance' which he incorporated in the allowance for disabilities.

Whilst observing that he would normally accept 5% as the proper rate of return, he preferred to acknowledge the 'new venture' by adjusting the rate of interest to 3.75% rather than include it as an end allowance under the heading of disabilities.

Tribunal decision		£
Total capital cost		1239000
less cost of excess runway	103070	
proportion of capital cost attributable to let portion	145000	248070
		990930
less to reflect 'tone of list' 12.5%		123866
		867064
less for disabilities (incl 'new venture' allce)		67064
		800000
Effective annual value at 3.75%		30000

Commentary
The case rehearses the arguments for and against the use of the profits method and concluded that it was inappropriate in this instance. In adopting the contractor's method, there were allowances in respect of excess capacity and the 'new venture' nature of the undertaking (the latter resulting in a 25% reduction in the effective annual value).

Example 14.5

Public school (Governors of Shrewsbury School v. Hudd (VO))
In an appeal against the gross value placed on the school, the rate-

payers' valuer built up his valuation by taking a unit price per 'equivalent boarder' (2.5 day boys being equivalent to one boarder); extra amounts were then added to the previous total for amenities not likely to be present in all public schools.

The valuation officer preferred to proceed by formulating a valuation using the contractor's basis. The valuation officer found an effective capital value as follows.

		£
Cost of modern substituted building		1 205 955
Average deduction 60.5%		730 526
		475 429
Add site works	47543	
professional fees	41838	
land	48980	138 361
Effective capital value		613 790

The Tribunal (Mr Erskine Sims and Mr H.P. Hobbs) felt that the valuation officer's starting figure was too high and that in arriving at his deduction he relied too much upon 'guidelines produced and too little upon his own judgment'.

They increased the allowance to 70% taking 3.5% to show a gross value of £15 750 for the main hereditament and £17 500 after adding in an agreed amount for outside houses.

The Tribunal went on to make other adjustments which are not relevant to this account.

Commentary

The Tribunal did not accept the unit price per boarder approach, preferring the contractor's basis approach of the valuation officer. The cost of the modern substituted building was reduced by a massive 70% in building up the effective capital value.

Example 14.6

A school within the University of London (*Imperial College of Science and Technology* v. *Ebdon (VO) and Westminster City Council*).

The valuation officer and the valuers for the other two parties all gave evidence. There was agreement that the contractor's basis was the most appropriate approach in ascertaining the assessment. There was only one difference of opinion as to the estimated replacement cost (ERC) of the 15 buildings involved, disagreement centring on the amount of the disability allowance. The parties ranged in their opinions from 0 to 54%.

Table 14.3 Comparison of allowances to arrive at the decapitalization rate

	Mr Eve %		Mr Ebdon %		Mr Hampsher %	
Minimum lending rate	5.5	Average of 20 yrs to end of 1972	*6.5	Average of 3¼ yrs to April 1 1973	(7.5)	at Oct 27 1972
						Inflation rate Dec 1972
Inflation	−4.0	Average inflation rate same period as above 4.08%	−3.5 or less	Average inflation rate 3 yrs to end of 1972 8.2%	−1.5	7.7%
Real rate of interest	1.5		3.0 or more		6.0	
Borrower's premium	+1.0		+2.0		+(0.5)-	
	2.5		5.0 or more		6.5	*Starting figure 8% last qtr 1972
Other factors	—		−3.5 or less		−1.5	
Long-term requirement	—		−2.5		—	
	2.5		2.5		5.0	
Depreciation Repairs	+0.75 +0.25		— +1.0		— +1.0	
	3.5		3.5		6.0	

* Valuer's starting point

The member of the Lands Tribunal (Mr C.R. Mallett) summarized the evidence given as to the appropriate allowances shown as Table 14.3 and then proceeded to use his discretion based on that evidence to determine the gross value for rating (Table 14.4).

The Tribunal discussed its decision through the five stages suggested in *Gilmore (VO)* v. *Baker-Carr* and added a sixth stage aimed at enabling the figures produced by the process to be reviewed. The member said

Table 14.4 Summary

Agreed estimated replacement cost of buildings	£20 340 438
Adjusted costs (see Appendix III) (−11.0%)	18 099 355
Effective capital value of land	
15.725 acres @ £400,000 *less* 11.0%	5 598 100
	23 697 455
Convert to annual rental value 3.5%	829 411
Less for disabilities −7½%	62 206
	767 205
Gross value, say	767 000

'. . . I think it is necessary and logical first to determine the annual equivalent of the likely capital cost to the hypothetical tenant on the assumption that he sought to provide his own premises, then to consider if this figure is likely to be pushed up or down in the negotiations between a hypothetical landlord and a hypothetical tenant having regard to the relative bargaining strengths of the parties.'

The result of these final deliberations is summarized in Table 14.5 which shows the make up of the final decision.

Further arguments relating to the decapitalization rate are reviewed in Chapter 13. Note also the text reference to decapitalization rates relating to educational establishments which have been fixed by regulation for assessments of rateable value for the National Non-Domestic Rate taking effect from April 1990.

The Lands Tribunal decision was upheld on appeal to the Court of Appeal; leave to appeal to the House of Lords was refused.

Commentary
Although the parties agreed that the contractor's basis was the correct approach to the valuation of the buildings, there was a wide divergence of view as to the amount of the disability allowance, if any.

Example 14.7
Purpose-built high security cash centre (*Barclays Bank PLC* v. *Gerdes (VO)*).

The building was of a warehouse type erected in 1982 but it differed from normal warehouses in that it was built to provide a secure building as a collecting, sorting, storing and distribution point for notes and coin.

Although the assessment of 15 similar properties throughout England were presented in evidence by the ratepayers, the valuation officer found them of no assistance as he could detect no consistent pattern of values. He therefore valued the premises by reference to the cost of construction using the contractor's basis of valuation.

Table 14.5 The six stages

1. Cost of construction Replacement or modern substitute	20 340 438
2. Effective capital value Allow deductions for age, obsolescence and other factors 11%	18 099 355
3. Land value limited to existing use £400 000 per acre less 11%	5 598 100
	23 697 455
4. Decapitalization rate based on 'real' rate of interest but allowing for a borrower's premium, 3.5%. (see text)	829 411
5. General considerations: disabilities taking account of items not previously considered 7.5%	62 206
	767 205
6. Negotiations Reflect the relative bargaining strengths of the parties	
Gross value, say	£767 000

The following valuation was presented to the Lands Tribunal by the valuation officer.

Valuation by P.J. Gerdes, ARICS (Valuation Officer)
Contractor's basis valuation

		(Figures presented to LVC)
1. Net cost of construction	£788 854*	(£788 854)
2. Apply indexation factor of 25% for 'tone' cost of construction	0.25	(0.25)
3. Net 'toned' capital value of building and site works	197 213	(197 213)
4. *Add* professional fees (+12%)	23 665	(23 665)
5. Gross effective capital value of building and site works at April 1 1973	220 878	(220 878)
6. Superfluity on specification	Nil −20%	(44 176)

7. Net adjusted effective capital value (excluding the land element)	220 878	(176 702)
8. *Add* 'toned' site value	25 000	(25 000)
9. 'Toned' effective capital value of land and building	245 878	(201 702)
10. Decapitalize @ 6% to gross value	0.06	(0.06)
	£14 752	(£12 102)
Gross value, say	£14 750	(£12 000)

* Source: Barclays Bank plc/E Dudley Smith & Partners (chartered quantity surveyors)

Valuation in accordance with sections 19 and 20 of the General Rate Act 1967, as at the date of the proposal: March 29 1983.

Two points will assist in understanding the figures. First, because the figures were based on a known cost of construction in 1982, it was necessary to reduce the figure to reflect the 'tone' of the valuation list on 1 April 1973. Second, there is a reference to 'superfluity' but no allowance. Originally the valuation officer had made a deduction for this item of 20% of the effective capital value of the building in 1973 and agreed it with the valuer for the ratepayer but he later decided that the deduction was not warranted. The allowance related to the need to install air-conditioning throughout the building because of the lack of windows. The failure to agree was rooted in the valuation officer's view that the building cost was approximately three times that of a standard warehouse unit, whilst the valuer for the ratepayers was contending for a gross value assessment less than 5% in excess of the assessment of a standard unit. The member of the Tribunal rejected the use of the contractor's basis in this case and criticized it on two counts. He said

'I do not think that the conversion of an actual 1982–3 cost to an estimated 1973 cost can be made with any precision by reference to the change in the estimates of approximate general building costs over the years as extracted from various publications on building cost estimates. This must be particularly so when dealing with a type of building that did not exist in 1973.

Second, in trying to arrive at the appropriate percentage in order to reduce capital value to annual value, I do not think that in the present case any proper consideration was given to the essential difference between capital cost and annual value in the case of adjoining standard warehouse units built by a developer on borrowed capital for speculative sale or letting, and in the case of the appeal premises, which were built for a bank for their own occupation, without any profit motive and with, presumably, an ability to make capital available on more favourable terms.'

Commentary

The decision demonstrates the difficulty of assessing a modern building for rating purposes, where the use was for specialized high security purposes although the unit gave the general appearance of a warehouse.

Example 14.8

You act for the directors of a private hospital service devoted to the treatment of patients suffering from terminal illness.

They occupy a Georgian house in extensive grounds on the outskirts of a large city. The original floor area has been doubled by the addition of single storey extensions over a period of years.

You have received instructions to prepare an asset valuation for incorporation into the annual accounts.

		£
Cost of modern substitute building having same gross internal floor area	1 640 000	
Fees, finance charges	474 000	
	2 114 000	
deduction for obsolescence	951 300	1 162 700
Open market value of land restricted to existing use (inclusive of costs)		250 000
Depreciated replacement cost		1 412 700

(subject to the adequate potential profitability of the business compared with the value of the total assets employed.)

Commentary

The building is of a type which would be difficult to value by reference to the open market. It comprises an old, converted building with modern additions and appears to have more land than necessary for its present purpose. There may be an opportunity of obtaining permission to develop the whole or part of the property.

However, for the present purpose, a valuation of the premises to include in the annual accounts, these issues are irrelevant. The premises are to continue to be used as a private hospital.

The valuer has assumed a modern substitute building for which the cost can be readily obtained and which is then discounted, heavily in this case, to reflect the relative unsuitability and inconvenience of the present set of buildings. A land value is then added, to arrive at the depreciated replacement cost. The value assigned to the land is limited to its value for the existing use.

The final figure is one which the directors should consider and adjust downwards where it is felt that the adequate profitability of the business is insufficient to support the use of the land and buildings asset.

A | **Appendix**

RATIOS USED IN THE INTERPRETATION OF BUSINESS ACCOUNTS

Profitability ratios

Return on capital employed (ROCE)
$$\frac{\text{profit before interest and tax}}{\text{total net assets}}$$

$$\frac{\text{gross profit}}{\text{sales}} \times 100$$

$$\frac{\text{net profit}}{\text{sales}} \times 100$$

Asset utilization
$$\frac{\text{sales}}{\text{fixed assets}}$$

$$\frac{\text{sales}}{\text{current assets}}$$

Liquidity ratios

Current
$$\frac{\text{current assets}}{\text{current liabilities}}$$

Liquidity (acid test)
$$\frac{\text{current assets (excluding stock)}}{\text{current liabilities}}$$

Capital gearing

Gearing relates to long-term financial stability and may be defined as a ratio:

$$\frac{\text{preference share capital and debentures}}{\text{ordinary share capital and reserves}}$$

A company with a high level of gearing may restrict distribution of dividends to shareholders giving priority to reduction of gearing by repayment of some of the borrowed capital. A highly geared company represents greater risks to shareholders but if the strategy is successful the shareholders will benefit to a greater extent than had the borrowing not taken place.

Interest cover

Interest on loan stock (debentures) must be paid regardless of the profit-ability of the company. This ratio emphasizes the magnitude of the obligation by showing the number of times the interest is covered by the profit

$$\frac{\text{net profit} - (\text{before tax and interest})}{\text{interest payable}}$$

Asset and creditor levels

Three ratios concerned with current assets and current liabilities may prove helpful

Stock turnover ratio	$\dfrac{\text{cost of sales}}{0.5(\text{opening stock} + \text{closing stock})}$
Average period of credit	
allowed to debtors	$\dfrac{\text{closing trade debtors} \times 365}{\text{turnover}}$
allowed by suppliers	$\dfrac{\text{closing trade creditors} \times 365}{\text{purchasers}}$

With the exception of working capital, it is not possible to indicate ratio norms which are industry specific; for this reason the ratios will be useful only where information is available of similar sized companies in the same line of business.

It should be emphasized that accounting ratios are nothing more than indicators of performance. The information requires careful interpreta-tion based on a deep knowledge of the particular type of business under consideration.

B	# Appendix

EXTRACTS FROM GUIDANCE NOTES RELATING TO ASSET VALUATIONS*

GENERAL PRINCIPLES TO BE OBSERVED IN PREPARING ASSET VALUATIONS

1. *General*
1.1 *Application of the Statements of Asset Valuation Practice.* The Statements of Asset Valuation Practice (SAVP) and Information Papers (IP) in this Handbook apply to the valuation of all fixed assets, that is, land and buildings of all kinds (including residential and agricultural land and buildings), and plant and machinery (including furniture and other chattels as defined in paragraph 2.11(b) of this SAVP). There are also references to land and buildings and plant and machinery held as current assets.
1.2 The SAVPs and IPs apply to all valuations which will be, or may be, included or referred to in any public or published document. They apply whether or not the Valuer has endeavoured to contract out of responsibility to third parties (that is, parties other than those to whom the valuation was addressed or for whom it was intended) and whether or not there is any likelihood that third parties will place reliance on the valuation. Thus they apply to valuations for the following purposes, amongst others:
   (a) incorporation in Company Accounts and other financial statements which are subject to audit;
   (b) incorporation in Stock Exchange prospectuses and circulars (SAVP No 20);
   (c) takeovers and mergers under the City Code (SAVP No 21);
   (d) valuations for Pension and Superannuation Funds (SAVP No 22), under the Insurance Companies Acts (SAVP No 19), for Property Unit Trusts (SAVP No 26) and for unit linked property assets of Life Assurance Companies (SAVP No 25); and

* These notes are regularly updated

(e) security for loans, debenture issues and mortgages (SAVP No 17) (other than those noted in paragraph 1.3(c) and (d) below).

1.3 The SAVPs and IPs do *not* apply to valuations undertaken only for the private information of the client. In such cases, Valuers are strongly advised to make clear in their report that it is for the private information of the client and not suitable for publication nor to be disclosed to any third party without consent.

1.4 The explanations of terms used in this Handbook on page xxi should be referred to, as this provides guidance on what is meant by such terms as 'Directors' and 'undertaking' and similar terminology used frequently within this Handbook.

1.5 *Extent of the Valuer's Instructions.* The Valuer should always discuss with the client and his advisers (amongst other matters) the purpose of the valuation, the properties to be valued, the plant and machinery to be included with (or excluded from) the valuation of buildings, the date and basis of valuation appropriate to the stated purpose, the nature of the information to be provided by the client to the Valuer and the extent to which the Valuer is to rely upon that information.

1.6 The extent and nature of the Valuer's instructions are entirely a matter between the Valuer and his client so long as the valuation is to be used for the client's private purposes and is not intended for publication nor to be disclosed to any third party without consent (see paragraph 1.3 above). However, in the case of valuations to which these SAVPs and IPs apply (see paragraph 1.2 above), it is essential that the Valuer should, at the outset, draw to the attention of the client any provisions of the SAVPs or IPs which are contrary to the client's initial instructions and that any publication or reference thereto will need to identify this fact.

1.7 Sometimes the Valuer's normal investigatory procedures will be restricted by a client (for example, he may not be permitted to make a proper inspection) or the Valuer may be given definite instructions to adopt assumptions which he would not normally make. Before accepting such instructions the Valuer should refer to SAVP No 6 and SAVP No 7 and establish the likelihood of the valuation being published. If such a restricted valuation becomes available to the public there is always a danger of it being misquoted or misunderstood. In such cases it is important that the Valuer sets out clearly in his Valuation Certificate the precise assumptions he has made, the extent to which his normal procedures have been restricted, and makes it clear that this has been done with the agreement of the client.

1.8 The Valuer should exercise great caution before permitting such restricted valuations to be used for other than the originally agreed purposes. If a valuation prepared on this basis is passed to a third party, even though the Valuer may have endeavoured to contract out of any responsibilities to third parties (the Hedley Byrne principle), it may well be that the recipient or reader will not fully appreciate the restricted worth of the valuation. If there is a risk that such a valuation might be passed to third parties, particularly the public, it would be appropriate to decline the instructions rather than risk the possibility of misleading users of the valuation.

2. *Classes and Categories of Fixed Assets*

2.1 There are three principal classes of fixed assets, namely 'non-specialised' and 'specialised' properties and 'plant and machinery'. Non-specialised and specialised properties may be further categorised according to the purpose for which they are held by the owner. Examples of such categories are:

(a) Land and buildings owner-occupied for the purposes of the business (SAVP No 9);

(b) Land and buildings held as investments (SAVP No 10);

(c) Land and buildings held as trading stock and work in progress (SAVP No 11);

(d) Land and buildings fully equipped as an operational entity and valued having regard to trading potential (SAVP No 12);

(e) Land and buildings held for development (SAVP No 13);

(f) Land and buildings in course of development (SAVP No 14);

(g) Land and buildings classified as a wasting asset (e.g. mineral-bearing land) (SAVP No 15);

(h) Land and buildings surplus to the requirements of the business (see paragraph 2.16 below).

2.2 Since different considerations, or bases of valuation, may apply, it is important to establish into which class or category any particular asset falls. The Valuer should offer advice, but it is ultimately the responsibility of the Directors of the undertaking to decide the class or category which applies to a property or group of properties. The Valuer should discuss classification or categorisation with the client if he considers the class or category decided is not appropriate.

2.3 If the Directors of the undertaking should decide to classify as 'specialised' a property which the Valuer considers should be 'non-specialised', the Valuer should include in the Valuation Certificate a statement to that effect and should indicate whether he thinks that the Open Market Value for the Existing Use would be likely to be substantially different from the reported amount.

2.4 *Non-Specialised Properties* are those for which there is a general demand, with or without adaptation, and which are commonly bought, sold or leased on the open market for their existing or a similar use, either with vacant possession for single occupation, or (whether tenanted or vacant) as an investment or for development.

AMONGST THE PRINCIPAL MATTERS CONSIDERED FOR A VALUATION ARE THE FOLLOWING

1. *Referencing*
   The following principal matters should be considered by the Valuer in his procedures for an asset valuation:
1.1 Characteristics of locality, availability of communications and facilities affecting value.
1.2 Age, description, use, accommodation, construction of any building, its installations, amenities and services.
1.3 Dimensions and areas of the land and buildings.
1.4 State of repair and condition.
1.5 Site stability (including the effects of mining and quarrying).

2. *Nature of Interest*
2.1 Tenure with reference to relevant restrictions, terms of leases (if leasehold), easements.
2.2 If owner-occupied.
2.3 Details of lettings and other occupations.

3. *Planning and Statutory Requirements*
3.1 Results of town planning, highways and other enquiries.
3.2 Contravention of any statutory requirements.
3.3 Outstanding statutory notices.

4. *Other Factors*
4.1 Rating assessments and other outgoings.
4.2 Any plant and machinery which would normally form an integral part of the building and, therefore, is included in the valuation.
4.3 Absence or otherwise of deleterious or hazardous materials.
4.4 Allowances for disrepair.
4.5 Any development potential.

5. *Market Analysis*
5.1 Details of comparable market transactions for either existing use or alternative use(s).
5.2 Market conditions and trends.
5.3 If valued on a depreciated replacement cost basis, the appropriate factors to be used.

6. *The Valuation Report or Valuation Certificate*

6.1 The presentation of the Valuation Report or Certificate will have regard to the need for any special format (e.g. the International Stock Exchange 'Admission of Securities to Listing' (the 'Yellow Book')) but will usually include a summary of the matters referred to above, and should include references to:

(a) the nature of the instructions and purpose of the valuation;

(b) the basis of the valuation;

(c) the date of the valuation;

(d) if with vacant possession (whole or in part);

(e) any assumptions;

(f) any special assumptions, together with additional valuations, if any;

(g) caveats as to:

    (i) structure, dry rot and the presence of deleterious materials (exclusion of a structural survey);

    (ii) non-publication;

    (iii) responsibility to third parties; and

    (iv) the sources of information and the need for verification;

(h) the value in figures and words; and

(i) signature of Valuer.

THE DEPRECIATED REPLACEMENT COST BASIS OF VALUATION

Statement

1. 'Depreciated Replacement Cost' (DRC) as a basis of valuation is the sum of:

(a) the Open Market Value of the land for its existing use; plus

(b) the current gross replacement cost of the buildings and their site works less an allowance for all appropriate factors such as age, condition, functional and environmental obsolescence which result in the existing property being worth less than a new replacement.

Guidance notes

2. The Valuer's attention is drawn to SAVP No 24 and IP No 14 which cover the valuation of public sector property assets.

3. The Depreciated Replacement Cost basis of valuation is used to arrive at the value to the business of those kinds of properties for which there is no market for a single owner-occupation and thus are rarely, if ever, sold in the open market for their existing use except

by way of a sale of the business in occupation due to the specialised nature of the buildings, their construction, arrangement, size, location or otherwise. Examples of the type of properties to which this basis applies are:

(a) Oil refineries and chemical works where usually the buildings are no more than structures or cladding for highly specialised plant.

(b) Power stations and dock installations where the buildings and site engineering works are related directly to the business of the owner and it is highly unlikely that they would have a value to anyone other than a company acquiring the undertaking.

(c) Properties located in particular geographical areas for special reasons or of such a size, design or arrangement as would make it impossible for the Valuer to arrive at a conclusion as to value from the evidence of open market transactions.

4. With regard to the land there may be a difference between value for the existing use and the value of undeveloped or virgin land. Many properties which fall to be valued on a depreciated replacement cost basis include large areas of land, in some instances extending to several hundred acres, as with a large oil refinery or chemical works. If the land were to be looked upon as virgin site being offered for sale in the open market, the price that would be obtained would allow for the fact that it may take the purchaser many years to carry out the development. This obviously does not arise in the context of land which is already fully developed. It is, therefore, necessary when putting a value on the land, to have regard to the manner in which it is developed by the existing building and site works and the extent to which these realise its full potential value. Incidental acquisition costs where material should be added to the amount of the value of the land.

5. The gross replacement cost of a building should take account of the following considerations:

5.1 A modern substitute building of the same gross internal floor area might cost substantially less than an identical replacement of the one on the site by taking advantage of current building techniques and the use of modern materials. This substitute may also be markedly different in its operating costs and life. The lower cost figure should be adopted.

5.2 The Valuer is concerned not with what it would cost to erect a building in the future but rather what it would have cost if work had commenced at the appropriate time so as to have the building available for occupation at the valuation date.

5.3 Additions will usually need to be made to the estimated building cost not only for professional fees but for finance carrying charges, which may be quite substantial when dealing with large property which may take several years to develop. The Valuer must consult with the Directors and the Auditors as to whether any irrecoverable VAT should be included and if so this should also be stated.

5.4   The Valuer should make it clear that any grants from Government or other sources have been ignored in the valuation. The accounting treatment of such grants can vary and the fact that they may be claimable should be ignored by the Valuer; they should not, therefore, be used to reduce the gross replacement cost of the buildings.

6.   The current gross replacement cost of the building is only the first part of the calculation, as the Valuer must then consider the deductions to be made to allow for the quality of the property as existing. Deductions will normally be made under three main headings:

6.1   *Economic obsolescence* — the age and condition of the existing building and the probable cost of future maintenance as compared with that of a modern building.

6.2   *Functional obsolescence* — suitability for the present use and the prospect of its continuance or use for some other purpose by the business. For example, a building constructed or adapted for specialised uses, including particular industrial processes, may have an apparent useful life longer than that contemplated for the actual operation carried on. The Valuer should always consult with the Directors to ascertain their future plans.

6.3   *Environmental factors* — existing uses should be considered in relation to the surrounding area and local and national planning policies.

7.   The Valuer must qualify every valuation prepared on a depreciated replacement cost basis as being subject to the adequate potential profitability of the business compared with the value of the total assets employed. It is for the Directors to decide if the business is sufficiently profitable to be able to carry the property in the balance sheet at the full depreciated replacement cost or whether some lower figure should be adopted. In the case of leasehold land the Valuer will need to draw attention to the amount of rent payable both in the present and future and any unusual onerous covenants which could affect the Directors' judgement on the adequacy of profits.

8.   Where a property is valued on a depreciated replacement cost basis, the amount of the valuation must be reported separately as this basis is shown separately from open market valuations in the accounts under 'Land and Buildings'.

9.   The Directors, in consultation with the Valuer, will need to make an assessment of the remaining economic useful life of the buildings, and this is discussed in more detail in SAVP No 18 on Accounting for Depreciation.

ACCOUNTING FOR DEPRECIATION

Statement

1. Statements of Standard Accounting Practice No 12 'Accounting for Depreciation' and No 19 entitled 'Accounting for Investment Properties' deal with depreciation in the context of the historical cost convention of accounting and in circumstances where fixed assets have been revalued in financial statements.

2.1 Depreciation is defined as the measure of the wearing out, consumption or other reduction in the useful economic life of a fixed asset whether arising from use, effluxion of time or obsolescence through technological or market changes.

2.2 SSAP 12 requires that for all financial statements (except in relation to investment properties) the depreciation of buildings and, in certain cases, land shall be allocated so as to charge a fair proportion to each accounting period of the estimated amount of the asset consumed during the expected useful economic life of the asset. All buildings have a limited life due to technological and environmental changes and they should be depreciated on the basis of their remaining useful economic life to the business. This requires an assessment of the future useful economic life of the buildings and an estimate of their cost or value to the business at the relevant date and is known as 'the depreciable amount'.

2.3 SSAP 19 provides that no periodic charges for depreciation are required for investment properties except for those held on lease, when the unexpired term is 20 years or less.

2.4 It is accepted that in normal circumstances depreciation is not applicable to freehold land. The exceptions include land which has a limited life due to depletion, for example, by the extraction of minerals, or which will be subject to a future reduction in value due to other circumstances, such as would arise where the present use is authorised by a planning premission for a limited period, following which it would be necessary to revert to a less valuable use.

2.5 Leasehold land and leasehold land and buildings must by their nature have a limited life to the lessee although the unexpired term of a lease may exceed the life of the buildings on the land. Regard must also be paid to any contractual or statutory rights to review the rent or determine or extend a lease.

2.6 The assessment of the depreciable amount and the remaining useful economic life of the asset are the responsibility of the Directors of the company or their equivalent in other organisations but it can be expected that Valuers may be consulted on these matters or on factors which are relevant to their assessment, e.g. degree of obsolescence, condition, market factors, town planning, etc.

3. *The Depreciable Amount*
    Assets may be stated in financial statements under historical cost
    accounting on one of the following bases:
    (a) Actual cost when acquired with or without subsequent deprecia-
        tion. This could be either:
        (i) the price paid for a completed property plus purchasing
            costs and depending on the accounting policy any works of
            improvement or adaptation; or
        (ii) the cost of the land and the cost of erecting the building;
    (b) a professional open market valuation made in a previous year
        with or without subsequent depreciation; or
    (c) a current professional open market valuation.

Guidance Notes

4. *Future Useful Economic Life*
4.1 In order to form an opinion as to the future useful economic life of
    buildings the Valuer will need to take into account the following
    matters:
    (a) *Physical obsolescence* — the age, condition and probable costs
        of future maintenance;
    (b) *Functional obsolescence* — suitability for the present use and
        the prospect of its continuance or use for some other purpose by
        the business. In the case of buildings constructed or adapted to
        meet the requirements of particular uses, including particular
        industrial processes, the Valuer will need to consult with the
        Directors to ascertain their future plans; and
    (c) *Environmental factors* — existing uses are considered in rela-
        tion to the present and future characteristics of the surrounding
        area, local and national planning policies and restrictions likely
        to be imposed by the planning authority on the continuation of
        these uses.
4.2 It is frequently difficult, if not impossible, to put a precise life on a
    building or group of buildings and Valuers may, therefore, have to
    resort to 'banding' of lives. Information should be available to
    identify buildings which are unlikely to remain beyond, say, 20
    years, and at the other extreme buildings with a life of more than,
    say, 50 years should be noted as having a life of 'not less than 50
    years'. It is apparent that the Valuer's task is made easier by the use
    of broad bands and in the majority of cases it is likely these will neet
    the company's requirements.
4.3 Where a property comprises a number of separate buildings, for
    example, large factory premises, it is suggested that in such cases
    the buildings should be grouped and, wherever possible, a single life

allocated to all buildings comprising the property. Such an approach can be justified by the fact that the life of individual buildings can usually be extended within reasonable limits by a higher standard of maintenance or minor improvement as it is normally uneconomic to carry out piecemeal redevelopment.

4.4   The allocation of a single life to all the buildings on a site would not be appropriate where parts of a property are used for different industrial processes which may give rise to changing accommodation requirements or where the company requires that each building must be considered individually.

5.   *The Depreciable Amount*

5.1   The Directors may seek advice from the Valuer and/or a number of other disciplines such as quantity surveyors or building surveyors where gross and net replacement costs of buildings are required to ascertain the depreciable amount.

5.2   Where the asset is shown in the balance sheet at a figure based on the original cost of the land and the cost of erecting the building, with or without subsequent depreciation, the cost of the building, less any subsequent depreciation which may have been taken, will be the depreciable amount.

5.3   Where the property has been acquired and is carried in the balance sheet at cost or has been the subject of a past or present open market valuation, which has been incorporated in the balance sheet, it is necessary for the Valuer to ascertain the value applicable to the buildings and the value of the land by an apportionment of the cost or the valuation as between buildings and land. The building element will be the 'depreciable amount' and the land element will be the 'residual amount' (termed 'residual value' in SSAP 12).

C | Appendix

EXTRACTS FROM THE RICS/ISVA CODE OF MEASURING PRACTICE

The purpose of the Code of Measuring Practice is to provide succinct and accurate measuring definitions for use when describing or specifying land and buildings, whether for conveyancing, planning, taxation, sale and letting particulars, valuation or other purposes.

The Code deals only with actual measurement practice; it does not specifically deal with certain types of building (such as licensed premises, holiday camps, or theatres), where special considerations may apply.

It is recommended that both metric and imperial units should be given.

General definitions

General

1. Gross External Area (GEA) (formerly sometimes referred to as 'Reduced Covered Area' or 'Gross Floor Space').

   Measurement of a building taking each floor into account to include:

   Perimeter wall thicknesses and projections;

   Areas occupied by internal walls and partitions;

   Columns, piers, chimney-breasts, stairwells and the like;

   Lift-rooms, plant rooms, tank-rooms, fuel stores whether or not above main roof level (except for Scotland, where for rating purposes these are excluded); and

   Open-sided covered areas and enclosed car-parking areas (should be stated separately).

   but excluding:

   Open balconies;

   Open fire escapes;

Open covered ways or minor canopies;

Open vehicle parking areas, terraces and the like;

Domestic outside WCs and coalhouses;

(the above exclusions should be separately calculated for building cost estimation purposes).

Party walls are to be measured to their central line.

Areas with a headroom of less than 1.5 m (5 feet) are excluded and quoted separately.

2. Gross Internal Area (GIA). Measurement of a building on the same basis as GEA, but excluding external wall thicknesses.

3. Net Internal Area (NIA) (formerly sometimes referred to as 'Effective Floor Area').

The usable space within a building measured to the internal finish of structural, external or party walls, taking each floor into account but excluding:

Toilets, toilet lobbies, bathrooms, cleaners' cupboards and the like;

Lift-rooms, plant-rooms, tank-rooms, other than those of a process nature, fuel stores and the like;

Staircases, lift-wells;

Those parts of entrance halls, landings and balconies used in common or for the purpose of essential access;

Corridors, where used in common with other occupiers or of a permanent essential nature (e.g. fire corridors, smoke lobbies, etc.);

High-tension and other areas under the control of supply or other external authorities;

Internal structural walls and partitions, pillars, stanchions, vertical ducts and the like;

The space occupied by permanent air-conditioning, heating or cooling apparatus and surface-mounted ducting which renders the space substantially unusable, having regard to the purpose for which it is to be used (where such apparatus is present its area may be quoted separately);

Car-parking areas; the number of car-parking spaces should be stated separately.

Floor space with a headroom of less than 1.5 m (5 feet) shall be excluded and may be quoted separately.

For shops: the depth to include recessed entrances, arcade displays, etc., in accordance with definition 19 below.

4. Gross External Cube (GEC). Measurements laterally as GEA multiplied by height. Vertical measurements are to be taken to the top of concrete foundations or 300 mm (1 foot) below the average ground level around the enclosing walls (whichever is the lower). The volume of the roof is determined by measuring:

(a) flat roofs to an assumed height of 600 mm (2 feet) above the top roof surface;

(b) pitched or lean-to roofs to half way up the height of the pitched roof. In no case, however, should the cubic content of the roof be less than it would have been had the rules for measurement of a flat roof been adopted.

5. Clear Height. The height between floor surface and lowest part of roof truss, ceiling beams, or roof beams at the eaves, excluding haunches.

6. Eaves Height.

(a) Internal. The height between the floor surface and the underside of the roof covering at eaves on internal wall face.

(b) External. The height between the ground surface and the underside of the roof covering at eaves on external wall face.

7. Ceiling Height. The height between the floor surface and the underside of the ceiling. (If a false ceiling is installed, the ceiling height to underside of structural ceiling may also be quoted.)

8. Site Area. The total area of the site within the site title boundaries, measured in a horizontal plane.

9. Gross Site Area. The site area, plus any area of adjoining roads, enclosed by extending the side boundaries of the site up to the centre of the road, or to 6 m (20 feet) out from the frontage, whichever is the less. (In compulsory purchase the latter limitation does not apply.)

10. Site Depth. The measurement of a site from front to rear boundaries (stating if forecourt is included).

11. Building Frontage. The measurement along the front of a building from the outside of external walls, or the centre line of party walls.

12. Site Frontage. The measurement of a site along its frontage between boundaries.

13. Plot Ratio. Ratio of the GEA to the Site Area where the Site Area is expressed as one, e.g. 3.5:1.

Shops

14. Sales Area. NIA usable for retailing purposes but excluding store-rooms unless they are formed by non-structural partitions, the existence of which should be stated if applicable.

15. Storage Area. NIA of that part of the building not falling under the definition of 'Sales Area' and usable only for storage.

16. Shop Frontage. The overall external frontage to shop premises (including entrance) but ignoring recesses, doorways and the like of other accommodation. Return shop frontage should be measured in like manner.

17. Overall Frontage. The gross measurement in a straight line across the front of the building, from the outside of external walls, or the centre

line of party walls. Return overall frontage should be measured in like manner.

18. Shop Width. Internal width between inside faces of external walls at front of shop or other point of reference. Unless otherwise stated, it will be assumed that the width is reasonably constant throughout the whole sales area.

19. Shop Depth. Measurement from back of pavement or forecourt to back of sales area. The existence of any non-structural partitions which should have been excluded, should be stated if applicable. Unless otherwise stated, it will be assumed that the depth is reasonably constant throughout the whole sales area.

20. Built Depth. Maximum external measurement from front to rear walls of a building at ground level.

B1 buildings

The joint working part (RICS/ISVA) considered the effect on the Code of the introduction of Class B1 under the Town and Country Planning (Use Classes) Order 1987.

It made the following observations and recommendations.

Class B1 embraces Business Use including offices (other than those for financial and professional services coming under Class A2), research and development and industrial processes subject to compatibility with the amenity to be expected in a residential area.

Thus the same building can qualify for either office or industrial use, which was not the case when the Code of Measuring Practice was published.

Buildings specifically constructed as offices, whilst having a notional capacity for alternative uses under B1 but not having the design facilities or market demand for wider B1 use, should continue to be measured on a Net Internal basis, as should all traditional office buildings. On the other hand, those buildings which are clearly designed to cater for the possibility of all categories of use within B1 should be measured on the basis Gross Internal Area.

This leaves buildings in the 'grey area' between the two obvious types referred to above and the Working Party feels that where there is any doubt, or comparisons have to be made for prospective tenants, both Gross Internal Area and Net Internal Area should be quoted.

D	# Appendix

BCIS: EXAMPLE OF COST ANALYSIS INFORMATION AVAILABLE

CI/SfB
847

DETAILED COST ANALYSIS

Students' Residence — 7 — a

BCIS Code: C — 3 — 1031

Job Title	Sainsbury Building (Hall of Residence), Worcester College, Oxford	Indices used to adjust costs to 1Q80 UK mean location base:
Location	Worcester College, Oxford	TPI at tender 212; at 1Q80 214 Location factor 0.98
Client	Provost and Fellows, Worcester College	

Date for receipt   March 1981   Base date: January 1981

INFORMATION ON TOTAL PROJECT

Project details
  Students residence for 30 students in three-storey stepped block.

Site conditions
  Sloping site, access through existing college buildings. New building is situated beside a lake.

Market conditions

Competitive Tender List

Tender documentation: Bill of Quantities

Selection of contractor: selected competition   787,747

Number of tenders — issued: 7
                         received: 7

Type of contract: JCT private contract 1963 edition

Cost fluctuations: NEDO Formula (full indices)

Contract period   stipulated by client: 18 months
                        offered by builder: 18 months
                        agreed: 18 months

ANALYSIS OF SINGLE BUILDING

Accommodation and design features
  Three-storey building, stepped to provide accommodation for students in groups of 6 rooms. Two groups of 6 rooms on ground and 1st floors, one group of 6 rooms and 2nd floor. Eighteen rooms have external terraces. Load bearing brick construction on concrete foundations bearing on gravel 2.7 m below ground. Concrete upper floors. Timber pitched roofs with natural slate coverings.

Areas   Functional units
  Basement floors   —   30 no. of persons
  Ground floor   —
  Upper floors   —
    Gross floor area   1031 m^2 Precentage of gross floor area

  Usable area   —
  Circulation area   —
  Ancillary area   —
  Internal divisions   —
    Gross floor area   1031 m^2

Floor space not enclosed   —
Internal cube   —   Storey Heights
                                                  Average   below ground floor   —
External wall area   —   at ground floor   —
Wall to floor ratio   —   above ground floor   —

BRIEF COST INFORMATION

Total contract
  Measured work   614651
  Provisional sums   36000
  Prime cost sums   40000
  Preliminaries   82096 — being 11.89 % of remainder of
  Contingencies   15000   contract sum (less contingencies)
  Contract sum   787747
Functional unit cost excluding external works — at tender date   22176.03 per no. of persons
                                              – at 1Q80, UK mean location   22841.98 per no. of persons

APPENDIX D

CI/SfB
847
.   .   .   .   .

Students' Residence — 7 — b

ELEMENT COSTS

Gross internal floor area: 1 031 m²   Base date: January 1981

Element		Preliminaries shown separately				Preliminaries apportioned		
		Total cost of element	Cost per m² gross floor area	Element unit quantity	Element unit rate	Total cost of element	Cost per m² gross floor area	Cost per m² at 1Q80, UK mean location
1	Substructure	80 836	78.41			90 445	87.73	90.36
2A	Frame	—	—			—	—	
2B	Upper floors	23 349	22.65			26 124	25.34	
2C	Roof	74 795	72.55			83 868	81.17	
2D	Stairs	2 151	2.09			2 407	2.33	
2E	External walls	59 779	57.98			66 885	64.87	
2F	Windows and external doors	51 265	49.72			57 359	55.63	
2G	Internal walls and partitions	45 653	44.28			51 080	49.54	
2H	Internal doors	39 531	38.34			44 230	42.90	
2	Superstructure	296 523	287.61			331 771	321.80	331.46
3A	Wall finishes	15 962	15.48			17 859	17.32	
3B	Floor finishes	21 552	20.90			24 114	23.39	
3C	Ceiling finishes	18 167	17.62			20 326	19.71	
3	Internal finishes	55 681	54.01			62 299	60.43	62.24
4	Fittings	25 394	24.63			28 412	27.56	28.38
5A	Sanitary appliances	5 000	4.85			5 594	5.43	
5B	Services equipment	—	—			—	—	
5C	Disposal installations	10 161	9.86			11 369	11.03	
5D	Water installations	included with element number 5E						
5E	Heat source	63 210	61.31			70 723	68.60	
5F	Space heating and air treatment	included with element number 5E						
5G	Ventilating systems	included with element number 5E						
5H	Electrical installations	32 388	31.41			36 238	35.15	
5I	Gas installations	75	0.07			84	0.08	
5J	Lift and conveyor installations	—	—			—	—	
5K	Protective installations	—	—			—	—	
5L	Communications installations	—	—			—	—	
5M	Special installations	—	—			—	—	
5N	Builder's work in connection	20 550	19.93			22 993	22.30	
5O	Builder's profit & attendance	4 784	4.64			5 353	5.19	
5	Services	136 168	132.07			152 354	147.77	152.00
	Building sub-total	594 602	576.72			665 281	645.28	664.65
6A	Site works	81 157	78.72			90 804	88.07	
6B	Drainage	14 892	14.44			16 662	16.16	
6C	External services	—	—			—	—	
6D	Minor building works	—	—			—	—	
6	External works	96 049	93.16			107 466	104.23	107.36
7	Preliminaries	82 096	79.63			—	—	
	Total (less contingencies)	772 747	749.51			772 747	749.51	772.01

Issued — BCIS 1983/1984 — 226 — 65

CI/SfB
847

SPECIFICATION AND DESIGN NOTES		Students' Residence — 7 — a

1	Substructure	RC ground beams on isolated pads. Concrete ground floor slab on poor bearing ground.
2B	Upper floors	RC upper floors with some downstand beams.
2C	Roof	Timber mono pitched roofs with natural slate coverings; overhanging eaves; lead cappings; verges etc. 100 mm roof insulation. Cast iron rainwater drainage.
2D	Stairs	RC stairs, timber handrails.
2E	External walls	Cavity walls with hand made facings externally, blockwork internally, and dry lining.
2F	Windows and external doors	Softwood windows and doors, stained, double glazing, aluminium sills.
2G	Internal walls and partitions	150 mm concrete block partitions, softwood copperlight glazed screens.
2H	Internal doors	Softwood doors and frames.
3A	Wall finishes	Plaster and emulsion to block walls. Tiling to bathrooms.
3B	Floor finishes	Carpet on screed to concrete floors. Vinyl flooring on screed.
3C	Ceiling finishes	Plywood, Nullifire finish on sloping soffits.
4	Fittings	Purpose made pine furniture: wardrobes, tables, beds, bookshelves. Folding blinds, pinboards, kitchen units etc.
5A	Sanitary appliances	Armitage Shanks lavatory basins, WCs, stainless steel sinks, steel baths, showers etc.
5C	Disposal installations	Soil and vent pipework.
5D	Water installations	Copper mains, hot and cold water installations.
5E	Heat source	Included with Element 5F.
5F	Space heating and air treatment	Low pressure hot water heating with pressed steel radiators. Individually operated booster system. Kitchen extract hoods.
5G	Ventilating systems	Included with Element 5F.
5H	Electrical installations	Normal electric power and lighting installations. Fluorescent lighting and spotlight fittings.
5N	Builder's work in connection	Forming chases and holes, ducts etc.
5O	Builder's profit and attendance	Normal profit and attendance on sub-contractors.
6A	Site works	Site preparation, landscaping and planting. GRP planting boxes with irrigation system. Hard landscaping including brick paviors. Random rubble wall.
6B	Drainage	Foul and surface water drainage.
7	Preliminaries	11.89% of remainder of contract sum (excluding contingencies).

Credits

Client	Worcester College, Oxford
Architect	MacCormac Jamieson & Prichard
Quantity surveyor	Hamilton H. Turner & Son
Consulting engineers	Buro Happold
General contractor	W.E. Chivers & Sons Ltd

Issued — BCIS 1983/84 — 226 — 66

Table A.1 BCIS concise cost analysis

Concise cost analysis

BCIS Code: A–2–765

				CI/SfB
				512.
Indices	214	276		Restaurants–1

Total project details (BCIS On-line Analysis No. 6142)

Job title: Restaurant and Bar		Measured work:	274 294
Location: Stratford upon Avon, Warwickshire		P.C. sums:	224 057
Client: Commercial Owner Occupier		Provisional sums:	3 750
Date for receipt of tender: 6 Nov 87	Date of Tender: 27 Oct 87	Preliminaries:	68 564
Contract period (months): stipulated 7	offered — agreed 7	Sub-total:	570 665
Type of contract: JCT private contract 1980 edition		Contingencies:	16 190
Fluctuations: firm		Contract sum:	586 855
Selection of contractor: selected competition			
Number of tenders: issued 7 received 6			

Analysis of single building

No. of storeys: 2
Gross floor area: 765 m²
Functional unit: 240 no. places

Element	Total cost of element	Cost per m²	Total cost of element inc. prelims.	Cost per m² inc. prelims	Cost per m² inc. prelims at 1Q80, UK mean location
Substructure	56 883	74.36	64 651	84.51	65.48
Superstructure	203 929	266.57	231 776	302.98	234.74
Internal finishes	36 990	48.35	42 041	54.96	42.58
Fittings	30 680	40.10	34 869	45.58	35.31
Services	96 320	125.91	109 473	143.10	110.87
Building sub-total	424 802	555.30	482 810	631.12	488.98
External works	77 299	101.04	87 855	114.84	88.98
Preliminaries	68 564	89.63	—	—	—
Total (less contingencies)	570 665	745.97	570 665	745.97	577.96

Type of construction: New two-storey Berni Inn restaurant. Concrete piled foundations, pile caps and ground beams, PCC ground, upper floor and stairs. Steel frame. Facing brick block cavity walls. Double glazed timber windows. Softwood pitched roof with plain clay tiling. Plaster and tiled walls, vinyl sheet tile and carpet floors, plasterboard and suspended ceilings some softwood boarding to walls and ceilings. Gas central heating and ventilation installation. (Burglar alarm, computer, refrigerator and telephone installations by client.)

Submitted by Leslie Clark and Partners
BCIS–1988/89–284–178

Further reading

Assets Valuation Standards Committee. (1990) Statements of Asset Valuation Practice and Guidance Notes. RICS.

Baum, A. and Crosby, N. (1988) Property Investment Appraisal. Routledge.

Butler, D. and Richmond, D. (1990) Advanced Valuation. Macmillan.

Davidson, A.W. (1989) Parry's Valuation and Investment Tables. Estates Gazette.

Emeny, P. and Wilks, H.M. (1984) Principles and Practice of Rating Valuation. Estates Gazette.

Fraser, W. (1984) Principles of Property Investment and Pricing. Macmillan.

Hager, D.P. and Lord, D.J. (1985) The Property Market, Valuations and Performance Measurement. Report of Proceedings, Institute of Actuaries.

McIntosh, A. and Sykes, S.G. (1985) A guide to Institutional Property Investment. Macmillan.

Reid, M. (1982) The Secondary Banking Crisis, 1973–1975, Macmillan.

Rose, J. (1979) Tables of the Constant Rent. The Technical Press.

Rose, J. (1981) The Construction of Valuation Tables. The Incorporated Society of Valuers and Auctioneers.

Salway, F. (1986) Depreciation of commercial property, College of Estate Management.

Trott, A. (ed). Property Valuation Methods: Research Reports. Interim, 1980, Final 1986. RICS.

Index of statutes and statutory instruments

STATUTES

STATUTORY INSTRUMENTS

Index of cases

Subject index